Overview of the Einstein-Podolsky-Rosen Argument with Applications in Physics, Chemistry and Biology

Pescara, Italy
2021

ISBN: 978-1-7138-6658-9

Printed from e-media with permission by:

Curran Associates, Inc.
57 Morehouse Lane
Red Hook, NY 12571

Some format issues inherent in the e-media version may also appear in this print version.

Published by Curran Associates, Inc. (2023)

For permission requests, please contact Hadronic Press
at the address below.

Hadronic Press
35246 US 19 North Suite 215
Palm Harbor, Florida 34684 USA

Phone: +1-727-940-3966

www.hadronicpress.com

Additional copies of this publication are available from:

Curran Associates, Inc.
57 Morehouse Lane
Red Hook, NY 12571 USA
Phone: 845-758-0400
Fax: 845-758-2633
Email: curran@proceedings.com
Web: www.proceedings.com

OVERVIEW
OF THE EINSTEIN-PODOLSKY-ROSEN ARGUMENT
with Applications in Physics, Chemistry and Biology

Ruggero Maria Santilli

Albert Einstein, Boris Podolsky and Nathan Rosen in the 1930s.

APAV - Accademia Piceno Aprutina dei Velati, Pescara, Italy
2021

C.F. 92036140678 - P. IVA 02184450688
CU Fatturazione elettronica: M5UXCR1
email: apavsegreteria@gmail.com
email PEC: apavsegreteria@pec.it

TABLE OF CONTENT

PREFACE

Jeremy Dunning-Davies

Almost from the moment it was first published in 1935, the famous - some might say infamous - article by Einstein, Podolsky and Rosen, claiming quantum mechanics to be an incomplete theory, has courted controversy. Following the initial furore, things calmed down with many accepting Bohr's refutation of the Einstein, Podolsky, Rosen argument. However, the topic has resurfaced periodically over the intervening years with no completely clear resolution emerging as far as some are concerned. In 2018 the entire issue resurfaced with the publication of some experimental results from a laboratory in Basle, results which served to support the view of that 1935 paper by Einstein, Podolsky and Rosen. This was followed by the international conference to be held in Florida in 2020 but which had to be turned into an online conference because of the covid-19 problem occurring in most, if not all, of the World. The conference ended up being a success at publicizing so many views, from so many people, which again supported the Einstein-Podolsky-Rosen line of thought. In particular, the conference brought to the fore the enormous, unheralded contribution to the debate by **Sir Ruggero Maria Santilli** (www.i-b-r.org/Ruggero-Maria-Santilli.htm - "Ruggero" hereon), with his first major contribution having come in 1998 after many years devoted to developing the new mathematics needed to cope adequately with the problems surrounding the Einstein, Podolsky, Rosen issue.

This book is intended to outline a collection of primary works from the mid 1960's to date by Ruggero and his collaborators on the verification and application of the Einstein-Podolsky-Rosen (EPR) argument that *Quantum mechanics* [and, therefore, quantum chemistry] *is not a complete theory,* in the expected recovering of classical determinism at least under limit conditions [1] (references indicated with square brackets refer hereon to those of Ruggero's Overview, while references indicated with upper numbers refer to additional works quoted in this preface).

This Preface is intended to provide a guide through a rather voluminous collection of works in various fields, as well as indicate important references following Ruggero's writing of the Overview.

During his Ph. D. studies at the University of Torino, Italy, in the mid 1960's, Ruggero discovered that the most advanced mathematics available at that time was insufficient for the representation of systems more complex than atomic structures, such as nuclear fusions, extended particles in deep mutual overlapping, combustion, biological structures, antimatter and other complex systems in the universe.

Extensive research carried out in 1964 at European mathematics libraries and repeated in 1978 at Cantabridgean mathematics libraries, convinced Ruggero that the mathematics needed for the effective representation of the indicated complex systems did not exist but had to be built and he had the courage to do it. In fact, Ruggero first recognized the need for new mathematics, constructed it himself and then proceeded to use it to examine complex systems not only in physics but in chemistry and biology as well.

In fact, throughout his research life, Ruggero first constructed the needed new mathematics and then passed to the treatment of complex systems in physics, chemistry or biology.

It then follows that no serious understanding of Ruggero's works is possible without a prior knowledge of the underlying new mathematics that, for the reader's convenience, are merely listed below with their primary references.

Quite notable is Ruggero's quote that "*There cannot exist a truly new physical theory without a new mathematics, and there cannot exist a new mathematics without new numbers.*"

I. NEW MATHEMATICS.

I-1. Lie-admissible mathematics.

As part of his Ph. D. curriculum, Ruggero studied the works by Giuseppe Luigi Lagrange who lived in Torino and wrote a number of papers in Italian. In this way, Ruggero learned that Lagrange represented physical reality via his celebrated analytic equations containing potentials represented with Lagrangeans L, plus *external terms* F representing non-conservative systems that, as such, are irreversible over time.

A comparison of Lagrange's original equations with the theories he was studying in Ph. D. courses soon revealed that Lagrange's external terms were not present in any of the available theories and that said external terms could not be represented with the available mathematics for various technical reasons.

A study of the works by Sir William Rowan Hamilton revealed their full parallelism with Lagrange's works, only formulated in phase space. In fact, the celebrated Hamilton's equations comprise a Hamiltonian H plus external terms F representing non-conservative and irreversible effects which were absent in the scientific literature of the mid 1960's.

Ruggero also learned that quantum mechanics cannot represent time irreversible systems because its main dynamical equations, Heisenberg's equation for an observable A, $idA/dt = [A, H] = AH - HA$ (where AH is the conventional associative product) can only represent the conservation of the energy, since $idH/dt = [H, H] = 0$.

To build the new mathematics needed for the representation of energy releasing, irreversible processes, Ruggero identified the main mathematical structure of quantum mechanics, which is given by Lie's theory with brackets $[A, B] = AB - BA$ and decided to generalize Lie's algebras into an algebra with brackets $(A, B) = A < B - B > A = ARB - BSA$ (where R, S are different, positive-definite operators) that turned out to be Lie-admissible according to the American mathematician, A. A. Albert [9]. The identification of the foundations of the new Lie-admissible mathematics immediately allowed Ruggero to generalize Heisenberg's equation into the form, today called *Heisenberg-Santilli Lie-admissible equations*, $idA/dt = (A, H) = A < H - H > A = ARH - HSA$ which represents the *time rate of variation* (rather than the conservation) of the energy $idH/dt = (H, H) = H(R-S)H \neq 0$ (see papers [6] [7] [8] of Ruggero's Ph. D. thesis although the unpublished version of the thesis better illustrates the originating thoughts).

The above studies in irreversibility attracted NASA attention and Ruggero moved to the U.S.A. in 1967 with his wife Carla and their newborn daughter Luisa for a one year appointment at the Center for Theoretical Physics of the University of Miami, Florida, with NASA support. He then accepted a faculty position at the Department of Physics of Boston University where he remained from 1968 to 1974 with partial support from the U. S. Air Force by teaching mathematics and physics at all levels and writing various papers with his associates and graduate students listed in his curriculum. From 1974 to 1977, Ruggero was a visiting scientist at the MIT Center for Theoretical Physics following an invitation from his Director Francis Low.

During his three year stay at the MIT-CTP, Ruggero wrote works[1−11] which are the *analytic foundation of verifications [210]-[214] of the EPR argument,* and comprise:

A. MIT-CTP preprints[1−5] on the necessary and sufficient conditions for the existence of a Lagrangian in field theory, technically known as the *conditions of variational selfadjointness* (SA), and their use for the representation of systems that are variationally non-selfadjoint (NSA) that are *analytic* in the sense of being derivable from a generalized action principle.

B. MIT-CTP memoirs[6,7,8] published at Annals of Physics which summarize the content of preprints[1−5].

C. MIT-CTP preprints[9,10] presenting the Newtonian particularization of the preceding field theoretical works, that Ruggero submitted in early 1977 to Springer-Verlag, Heidelberg, Germany, for publication as monographs.

D. MIT-CTP preprint[10] intended to be a third field theoretical volume of the Newtonian references[9,10], which preprint has remained unpublished, yet available as Annals of physics papers[6,7,8].

As we shall see in Section III, the above field theoretical works are important for the EPR completion of quantum electrodynamics into a form representing deviations of the theory from recent measures.

In September 1977, Ruggero joined Harvard University with a joint appointment as a visiting scientist at the Lyman Laboratory of Physics and at the Department of mathematics. In view of his works at MIT, Ruggero received on arrival an invitation from the U. S. Department of Energy for a grant intended to search for possible new clean nuclear energies. Soon thereafter, Ruggero also received the acceptance from Springer-Verlag for the publication of MIT preprints[9,10].

Encouraged by these openings, Ruggero plunged himself, firstly, in the construction of the Lie-admissible mathematics and secondly, in its application for the treatment of energy releasing irreversible systems.

The resulting main studies at Harvard University in the Lie-admissible mathematics, also called *genomathematics*[6,7,8], are given by: the 200 page memoir [19] of 1978 setting up the foundations of the new mathematics; Springer-Verlag monographs [21] [22] of 1978 proposing for the first time the completion of quantum mechanics into *hadronic mechanics* (see page 112) via the Lie-admissible generalization of Lie's theory and Heisenberg's equations and monographs [39] [40] on the Lie-admissble formulation of various aspects of 20th century applied mathematics.

In 1982, Ruggero accepted the position of President and Professor of Physics at the Institute for Basic Research (IBR) located at the Prescott

House within the compound of Harvard University, which was moved to Florida in 1989 (www.i-b-r.org).

Among a rather large scientific production at the IBR reviewed in the Overview, we here merely mention the systematic presentations of the various branches of hadronic mechanics in monographs [23]-[25] of 1996, and the five volumes [74] of 2006, all works having extensive references on independent contributions.

Various lectures on Lie-admissble mathematics are available on the website www.world-lecture-series.org. A tutoring lecture specifically intended to the verification of the EPR argument is available in Ref. [145].

Contributions in Lie-admissible mathematics that are important for the verification of the EPR argument are the following:

1. Paper [37] of 1979 established the bimodular structure of Lie-admissible mathematics, in the sense that the product $(A, B) = A < B - B < A = ARB - BSA$ can be reduced to two nodular actions, one to the right (representing motion forward in time) $H > |\psi >= HS|\psi >= E|\psi >$, and one to the left (representing motion backward in time) $< \psi| < H =< \psi|RH =< \psi|E', \quad E' \neq E$, whose in-equivalence assures the axiomatic representation of irreversibility. Said bimodular structure also assured the preservation of quantum mechanical axioms by the Lie- admissible branch of hadronic mechanics, since in a bimodular structure, quantum axioms are merely formulated per each selected time ordering.

2. In early 1980, it became known that physical applications of Lie-admissible methods are inconsistent when formulated over conventional numeric fields. This impasse was resolved with the discovery in the 1993 paper [33] of the genotopic numbers with multiplicative genounit to the right $I^{>} = 1/s$ and to the left ${}^{<}I = 1/R$.

3. In the mid 1990's it became also known that the representation of extended particles via Lie-admissible methods were inconsistent when

elaborated via Newton-Leibnitz differential calculus due to its strict local character. In particular, hadronic mechanics was still missing in the mid 1990's a consistent generalized formulation of the angular momentum due to the lack of a consistent generalization of the quantum mechanical linear momentum $p|\psi(r)>= -i\hbar\partial_r|\psi(r)>$. These additional impasses were resolved via the generalization-completion in Ref. [34] of 1996 of the conventional differential calculus into a form defined on *volumes*, rather than points, with *genomomentum* $p^> > |\psi^>(r^>)>= -i\partial^>_{r>}|\psi^>(r^>)>= -iI^>\partial_{r>}|\psi^>(r^>)>$ and to the left $<\psi^<(^<r)| <^< p = -i\partial^<_{<r} < \psi^<(^<r)| = -i^< I\partial_{<r}|\psi^<(^<r)>$ where volumes are represented by the *genounits* $I^> = 1/S$ and $^<I = 1/R$, respectively.

4. Paper [35] of 1989 identified a very simple method for the completion of 20th century mathematics into the Lie-admissible covering via the following two nonunitary transformations $UW^\dagger = I^>,\ \ WU^\dagger =^< I,\ \ UU^\dagger \neq I,\ \ WW^\dagger \neq I$ and the proof of the invariance over time of the genounits with consequential invariance of the shape and density of the represented particles.

5. Memoir [41] of 2006 proved the universality of the Heisenberg-Santilli genoequation f for the representation of all possible (regular) nonlinear, non-local, non-conservative, NSA systems via realizations of the type $R = 1,\ S = 1- = F/H$ with dynamical equation $idA/dt = (A, H) = A < H - H > A = AH - HA - AF$ where F is a suitably normalized operator form of Lagrange's and Hamilton's external terms.

In closing, we should indicate that there exists considerable literature on Lie-admissible algebras within the context of non-associative algebras in pure mathematics (see, e.g., Ref.[13] and Vol. I of Refs. [74]). However, these studies are formulated over conventional numeric fields, and even though mathematically correct, they cannot be used for the verification of the EPR argument because of a number of insufficiencies identified in Ruggero's Overview.

I-2. Lie-isotopic mathematics.

All objections against the EPR argument (see, e.g., Refs. [2]-[4]) are formulated for quantum mechanical, thus *conservative* systems for which Lie-admissible mathematics is inapplicable.

Additionally, in order to search for possible new clean nuclear energies under his DOE grant, Ruggero had to study nuclear structures, that when stable and isolated, verify all conventional total conservation laws, yet admit a more general formulation of internal strong interactions.

These requirements mandated the construction of a new mathematics for the representation of isolated, thus conservative systems with *extended* constituents in deep mutual penetration-entanglement, under non-linear, non-local and NSA interactions, yet such to verify conventional total conservation laws.

The needed new mathematics was identified in paper [19] of 1978 as the particular case of the Lie-admissible mathematics for $R = S = T > 0$ with basic brackets $[A, B]^* = A \star B - B \star A = ATB - BTA$, where T is called the isotopic element, which verify Lie's axioms, for which reason Ruggero called the new mathematics *Lie-isotopic* or isomathematics for short. Following the original proposal [19], isomathematics was studied in detail in monograph [22] of 1981 via the identification of its universal enveloping isoassociative algebra ξ with isoproduct $A \star B = ATB$, the initiation of the isofunctional analysis and the isotopic completion of the various branches of Lie's theory.

Isomathematics achieved maturity with the discovery of isonumbers [33], the isodifferential calculus [34] and the simple method for its construction in Ref. [35]. Systematic presentations of isomathematics were then provided in monographs [23]-[25] and [74].

Various lectures in isomathematics are available from www.world-lecture-series.org. A tutoring lecture in isomathematics specifically intended for the verification of the EPR argument is available from Ref. [143].

We should note the completion of the quantum entanglement of particles into the covering *EPR entanglement* (introduced by Ruggero in his Overview and first released in paper[18]), which represents the non-linear, non-local and NSA interactions due to the overlapping of the wavepackets of particles via realizations of the isotopic element T, by jointly providing an explicit and concrete realizations of Bohm's *hidden variables* [17] (see Figure 1 and Section 7.3.2 of the Overview).

There is little doubt that Ruggero's EPR entanglement will have applications in all quantitative sciences. To see it, it is sufficient to note that the conception of a nucleus, or a molecule or a virus, as being composed of extended constituents under EPR entanglement, implies the inapplicability of all objections against the EPR argument [2]-[4], thus opening the door for a new physics as well as chemistry and biology.

We should also indicate that isomathematics has attracted considerable interest in mathematical circles and has seen a number of important contributions by pure mathematicians identified in the Overview, with complete listing in Vol. I of monographs[74].

I-3. Hypermathematics.

In line with his belief that mathematics will never admit final formulations, Ruggero has stated various times, that despite their vast representational capabilities, Lie-admissible and Lie-isotopic formulations cannot describe "all elements of reality" [1] because they are single-valued (in the sense that the multiplication of two quantities yields one single result, e.g., $2 \times 3 = 6$). This is an excessive limitation for the representation of complex systems, such as biological structures, which suggested Ruggero to turn Lie-admissible and Lie- isotopic mathematics into *multi-valued* (rather than multi-dimensional) forms via genotopic $R,\ S$ and isotopic T elements representing an ordered set of values. As an example, the assumption that the isotopic element has three values $T = \{2, 3, 1/2\}$ implies that $2 \times 3 = \{12, 18, 1/3\}$ [27].

Even though sufficient for "simple" biological structures such as seashells, the above formulation of hypermathematics turned out to be insufficient for the representation of a living organism such as a cell.

Thanks to the participation of the Greek mathematician Thomas Vougiouklis, the above multi-valued formulations were generalized-completed into the most general and complex mathematics conceivable nowadays by the human mind, that of the *Lie-admissible and Lie-isotopic hyperstructures defined on hyperspaces over hyperfields* (see the Tutoring Lecture [146]).

I-4. Isodual mathematics.

Despite the advances indicated above, Ruggero still considered Lie-admissible and Lie-isotopic mathematics as being unable to represent *all* elements of reality [1].

During his graduate studies in the mid 1960's, Ruggero wanted to study whether a far away galaxy is made up of matter or of antimatter, but was prohibited from doing such a study because the most advanced mathematics and physics available at that time identified no difference between matter and other complex systems in the universe.

Additionally, Ruggero has been a supporter of Dirac's view that *antimatter has negative energy* [11], as a pre-requisite for the representation of matter-antimatter annihilation.

Recall that, as discovered by Dirac himself, negative energies violate causality in the sense that the effect generally precedes the cause in the solution of quantum mechanical equations with negative energy.

For the intent of resolving Dirac's causality problem, while being at the Department of mathematics of Harvard University in the early 1980's, Ruggero decided to build the foundations of yet another new mathematics, this time based on the *negative unit* "-1" under the name of *isodual mathematics* where the word "isodual" stands to indicate an axiom-preserving duality of 20th century mathematics.

This lead to the construction of the isodual images of the conventional, Lie-isotopic and Lie-admissible mathematics [29].

In view of the construction of the above new mathematics and their application in physics, chemistry and biology, Ruggero was listed in 1990 by the Estonia Academy of Sciences among the most illustrious applied mathematicians of all times (see http://santilli-foundation.org/santilli-nobel-nominations.htm).

II. VERIFICATIONS OF THE EPR ARGUMENT.

Thanks to the new mathematics indicated in the preceding section, Ruggero and his associates have constructed the axiom-preserving completion of quantum mechanics into *hadronic mechanics* comprising the *Lie-isotopic, Lie-admissible, hyperstructural and isodual branches* [23]-[25] [74], with corresponding completions for *hadronic chemistry* [30] and *hadronic biology* [27] (see the outline in Section 6 of the Overview). Following these preparatory studies, as well as the teaching from historical completions of quantum mechanics by W. K. Heisenberg [16], Prince L. V. P. R. de Broglie [17], D. J. Bohm [18] and others reviewed in Section 5 of the Overview, Ruggero achieved the following verifications of the EPR argument (see also the presentation at the 2020 EPR conference by E.T. D. Boney om Gödel's incompleteness theorems [217] and by A.A. Nassikas on the minimum contradiction theory [218], as well as their recorded talks [0]:

II-1. Verification of the EPR argument for irreversible processes.

We are here referring to Ruggero's 1967 Ph. D. thesis [6]-[8] (see also memoir [41]) in which he proved the lack of completion of quantum mechanics for energy releasing, thus time irreversible processes and constructed the foundations of the Lie-admissible completion of quantum mechanics, also called genomechanics.

Besides scientific and industrial applications, the irreversible character of the Lie-admissible mechanics has stimulated studies to achieve a connection between mechanics and thermodynamics, of course, via the intermediate step of irreversible statistical mechanics, see P. Roman et al[14], A. A. Bhalekar[15], J. Fronteau et al [42], J. Dunning Davies [52] and other contributions.

II-2. Verification of the EPR argument for classical counterparts.

In a paper of 1964, J. S. Bell [3] proved a theorem essentially stating that a system of quantum mechanical particles with spin $1/2$ does not admit a classical counterpart.

Thanks to the prior development of the Lie-isotopic mathematics, Ruggero proved in his 1998 paper [210] that Bell's theorem is inapplicable (rather than being violated) for a system of extended particles with spin $1/2$ under deep EPR entanglement and said system does indeed admit a fully defined classical counterpart.

The proof was essentially based on the completion of Bell's theorem via the isoproducts $A \star B = A\hat{t}B$, $\hat{T} > 0$, which allows an explicit and concrete realization of Bohm's *hidden variables* [17].

In paper [210], Ruggero also achieves a numerically exact representation of nuclear magnetic moments via hadronic mechanics that escaped a quantum mechanical representation for about one century.

II-3. Verification of the EPR argument for classical determinism.

In a paper of 1981 [47], Ruggero introduced a generalization-completion of Heisenberg's uncertainties for strong interactions among extended hadrons in deep mutual entanglement (as occurring in a nuclear structure) when they are represented via hadronic mechanics.

In the 2019 paper [211], Ruggero proved the progressive recovering of Einstein's determinism in the interior of hadrons, nuclei and stars and its full recovering at the limit of gravitational collapse.

The proof is based on the realization of the isolinear momentum via the isodifferential calculus [34], $\hat{p} \star |\hat{\psi}(\hat{r}) >= -i\hat{\partial}_{\hat{r}}|\hat{\psi}(\hat{r}) >= -i\hat{I}\partial_{\hat{r}}|\hat{\psi}(\hat{r}) >$, $\hat{I} = 1/\hat{T} > 0$ (where "hat" denotes definition in hadronic mechanics).

Let Heisenberg's uncertainties be given by $\Delta p \Delta r = (1/2) < \psi(r)|[r, p]$-$|\psi(r) > \leq (1/2)\hbar$ under the normalization $< \psi||\psi >= \hbar$. The isotopies lead uniquely and unambiguously to the *isouncertainties* $\Delta\hat{p}\Delta\hat{r} = (1/2) < \hat{\psi}(\hat{r})| \star [\hat{r}, \hat{p}] \star |\hat{\psi}(\hat{r}) > \leq (1/2)\hat{T}$ under the isonormalization $< \hat{\psi}| \star |\hat{\psi} >=< \hat{\psi}|\hat{T}|\hat{\psi} >= \hat{T}$. The results of paper [210] then follow from the fact that according to all fits of experimental data in hadron and nuclear physics, the isotopic element $\hat{T} = 1 - F/H$ has very small values and represents Schwartzchild's horizon at the limit of gravitational collapse.

II-4. Verification of the EPR argument for electron valence bonds.

In the final statement of their historic paper [1], Einstein, Podolsky and Rosen state that the *wavefunction of quantum mechanics* [and, therefore, of quantum chemistry] *cannot represent all elements of reality.*

In the 2001 monograph [30] on hadronic chemistry (see Chapter 4 on), Ruggero proved the above statement by showing that the wavefunction $\psi(r)$ of the Schrödinger equation of quantum chemistry cannot represent the *attraction* between the *identical* electrons in valence bonds since they experience at 10^{-13} *cm* the extremely big *repulsive* force of 230 *Newton.*

By subjecting the Schrödinger equation for electron valence bonds to a non-unitary transformation of the type $UU^\dagger = \hat{I} = exp\{-\bar{\psi}/\psi \int \psi^\dagger \psi d^3 r\}$, and by using an appropriate selection of $\bar{\psi}$, the repulsive Coulomb potential is transformed into a strongly attractive Hulten potential, with corresponding completion of the wavefunction into a form representing the *element of reality* given by electron valence bonds.

The resulting bound state was called *isoelectronium* and allowed the numerically exact representation of the experimental data for the hydrogen [31] and water [32] molecules that have escaped quantum chemistry for half a century.

II-5. Verification of the EPR argument for antimatter.

Ruggero did not accept the conventional charge conjugation $\psi(t, r) \rightarrow \psi^c(t, r) = -\psi^\dagger(t, r)$ as being final because it provides no conjugation of matter into antimatter for neutral particles and prevents a representation of matter-antimatter annihilation (because antiparticles have the same positive energy of particles).

Hence, Ruggero developed the isodual mathematics for antimatter [29] outlined in the preceding section which is characterized by the isodual map (indicated with an upper letter "d") applied to the totality of quantities and their operations of quantum mechanics, resulting in the new *isodual charge conjugation* $\psi(t, r) \rightarrow \psi^d(t^d, r^d) = -\psi^\dagger(-t^\dagger, -r^\dagger)$, under which antimatter has negative energy as predicted by Dirac [11], evolves backward in time and all its characteristics are opposite those of matter, thus allowing a representation of matter-antimatter annihilation.

The violation of causality for particles with negative energies is resolved by the isodual mathematics because *particles with negative energy or negative time referred to negative units are as causal as positive energies or positive time referred to positive units.*

III. NEW APPLICATIONS.

Section 8 of Ruggero's Overview and the vast literature quoted therein, illustrate quite clearly that the verifications of the EPR argument, the new notion of EPR entanglement and their embodiment in the completion of quantum into hadronic mechanics, chemistry and biology, have important *new* implications (that is, applications not permitted by quantum mechanics) in all quantitative sciences.

We here merely note that what appears to be a central implication of the EPR argument, the existence of superluminal speeds under strong interactions (Section 8.4.4-VI), was first voiced by Ruggero in 1982[16].

The exact character of quantum electrodynamics (QED) has been recently disproved by accurate measurements[12,17] establishing a deviation of the measured muon g-factor from the QED prediction. Following the release of his Overview, Ruggero has provided a numerical representation of the anomalous muon g-factor via a branch of hadronic mechanics called *IsoElectrodynamics* (IED)[18] whose analytic counterpart is given by the MIT-CTP papers[6,7,8].

We should finally mention that, following the release of the Overview, an independent review of the Einstein, Podolsky, Rosen argument has been published[19] and this is recommended for reading before embarking on a study of the technical issues as outlined in some detail in this book. In this article, inspired partly by the above-mentioned 2020 Florida Conference, much of the latest information, both experimental and theoretical, pertaining to this topic - a topic whose resolution is so important for the future of all science, not just physics - is provided. Also, reference is made to some earlier work supporting the Einstein-Podolsky-Rosen stance, which as far as many are concerned, has remained conveniently almost hidden ever since its ideas were advanced in the mid-1980's.

Further, the article contains some purely reflective thoughts on the position of probability theory in physics and other sciences, thoughts which may have relevance in other disciplines as well. The article also gives an independent view of some of the possible consequences of Ruggero's work - consequences which could affect each and every one of us, with the possibility of a new method for the quick and safe disposal of nuclear waste being probably the most important for many.

Even this, though, is merely one side product of his work which could lead to many more benefits for mankind.

In many ways, as with so many issues, the main problem encountered in discussions of the Einstein, Podolsky, Rosen issue has been an unwillingness to think 'outside the box', as the saying goes. However, the final resolution of this and other outstanding questions facing modern day science will surely rely on unorthodox thinking and in this respect, all should remember that final paragraph in the 1958 edition of Dirac's well-known text on quantum mechanics:

It would seem that we have followed as far as possible the path of logical development of the ideas of quantum mechanics as they are at present understood. The difficulties, being of a profound character, can be removed only by some drastic change in the foundations of the theory, probably a change as drastic as the passage from Bohr's orbit theory to the present quantum mechanics.

This powerful statement from such an eminent theoretical physicist surely deserves careful contemplation and certainly cannot be dismissed easily. It is the contention here that Ruggero has achieved, at least in part, that drastic change envisaged as necessary by Dirac. Not that Ruggero would claim this work constituted a complete theory, because as he has said on so many occasions, there cannot be a truly complete theory which successfully encompasses every possible situation. Physics, and indeed all of science, will always be continuously evolving subjects but it is felt the work discussed in this book represents a significant step forward in helping man gain a better understanding of the universe in which we all live.

Acknowledgments. The author would like to thank Ruggero Maria Santilli and the R. M. Santilli Foundation for continuous contacts and communications without which this Preface would not have been possible. Thanks are also due to all participants of the 2020 International Teleconference on the EPR Argument for comments and suggestions.

Jeremy Dunning-Davies
University of Hull,
Hull, England.
Email: masjd@masjd.karoo.co.uk

May 31, 2021

REFERENCES

1. R. M. Santilli, "Necessary and sufficient conditions for the existence of a Lagrangian in field theory,"
Part I: "Variational approach to selfadjointness for tensorial field equations," MIT-CTP preprint no. 533 (1976).

2. R. M. Santilli, "Necessary and sufficient conditions for the existence of a Lagrangian in field theory,"
Part II: "Direct Analytic Representations of Tensorial Field Equations," MIT-CTP preprint no. 534 (1976).

3. R. M. Santilli, "Necessary and sufficient conditions for the existence of a Lagrangian in field theory,"
Part III: "Generalized analytic representations of tensorial field equations," MIT-CTP preprint no. 560 (1976).

4. R. M. Santilli, "Necessary and sufficient conditions for the existence of a Lagrangian in field theory,"
Part IV: "Isotopic and genotopic transformations of the lagrangian for tensorial field equations," MIT-CTP preprint no. 609 (1976).

5. R. M. Santilli, "Necessary and sufficient conditions for the existence of a Lagrangian in field theory,"
Part V: "The problem of symmetries and conservation laws for tensorial field equations," MIT-CTP preprint 610 (1976).

6. R. M. Santilli, "Necessary and sufficient conditions for the existence of a Lagrangian in field theory,"
Part I: Variational approach to self-adjointness for tensorial field equations
Annals of Physics, **163**, 354-408 (1977),
http://www.santilli-foundation.org/docs/Santilli-45.pdf

7. R. M. Santilli, "Necessary and sufficient conditions for the existence of a Lagrangian in field theory,"
Part II: Direct analytic representations of tensorial field equations
Annals of Physics, **163**, 409-468 (1977),
http://www.santilli-foundation.org/docs/Santilli-46.pdf

8. R. M. Santilli, "Necessary and sufficient conditions for the existence of a Lagrangian in field theory,"
Part III: Generalized analytic representations of tensorial field equations,
Annals of Physics, **165**, 227-258 (1977),
http://www.santilli-foundation.org/docs/Santilli-47.pdf

9. R. M. Santilli, *Foundations of Theoretical Physics*, Vol. I: *The Inverse Problem in Newtonian Mechanics*, MIT-CTP no. 611-1, 389 pages (1977).

10. R. M. Santilli, *Foundations of Theoretical Physics*, Vol. II: *Generalization of the Inverse Problem in Newtonian Mechanics*, MIT-CTP no. 611-2, 290 pages (1977).

11. R. M. Santilli, *The Inverse Problem in Field Theory: Lagrangian Representations of Lorentz Covariant Field Equations with Arbitrary Couplings*, MIT-CTP no. 612, 389 pages (1977).

12. James P. Miller, Eduardo de Rafael and B. Lee Roberts, "Muon (g-2): experiment and theory," Rep. Prog. Phys. **70**, 795-881, (2007),
http://g2pc1.bu.edu/ roberts/Miller-dR-R-rpp7_5_R03.pdf

13. M. L. Tomber, "A short history of non-associative algebras," Hadronic J. **2**, 1252-1387 (1979).

14. P. Roman and R. M. Santilli, "A Lie-admissible model for dissipative plasma," Lettere Nuovo Cimento, Vol. 2, 449-455 (1969).

15. A. A . Bhalekar, "Santilli's Lie-Admissible Mechanics. The Only Op-

tion Commensurate with Irreversibility and Nonequilibrium Thermodynamics," AIP Conference Proceedings 1558, 702 (2013); doi: 10.1063/1.48-25588,
http://www.santilli-foundation.org/docs/bhalekar-lie-admissible.pdf

16. R. M. Santilli, "Can strong interactions accelerate particles faster than the speed of light?" Lettere Nuovo Cimento **33**, 145 (1982),
www.santilli-foundation.org/docs/Santilli-102.pdf

17. Fermilab release: muon g-2 experiment,
https://muon-g-2.fnal.gov/

18. R. M. Santilli, "Representation of the anomalous magnetic moment of the muons via the Einstein-Podolsky-Rosen completion of quantum into hadronic mechanics,"Progress in physics, Vol. 17, 210-215 (2021),
http://www.santilli-foundation.org/muon-anomaly-pp.pdf

19. J. Dunning-Davies, "A Present Day Perspective on Einstein-Podolsky-Rosen and its Consequences," Journal of Modern Physics, **12**, 887-936 (2021),
https://www.scirp.org/journal/paperinformation.aspx?paperid=109219

OVERVIEW
OF THE EINSTEIN-PODOLSKY-ROSEN ARGUMENT
with Applications in Physics, Chemistry and Biology

Ruggero Maria Santilli

ABSTRACT

In the 2020 Teleconference Ref. [0]: 1) We studied the 1935 objections against the **quantum entanglement** moved by Einstein, Podolsky and Rosen (EPR argument) [1]; 2) We pointed out that the interactions caused by wave-overlapping are of contact, zero-range, non-linear, non-local and non-Hamiltonian type; 3) We studied the new type of particle entanglement, here called **EPR entanglement**, consisting of particles in continuous and instantaneous communication via the overlapping of their wavepackets, thus without any need for superluminal speeds [1], whose non-Hamiltonian interactions are represented by the isotopic element $\hat{T}$ in the axiom-preserving product $A\hat{\times}B = A\hat{T}B$ of isomathematics and related hadronic mechanics (Tutoring Lectures [143] [144]); 4) We showed that the isotopic element $\hat{T}$ provides an explicit and concrete realization of Bohm's hidden variables; 5) We reviewed the recent verifications of the EPR argument by R. M. Santilli [210]-[214] showing the inapplicability of Bell's inequalities and other objections against the EPR argument for extended particles under non-potential interactions, with ensuing progressive recovering of classical images and Einstein's determinism in the structure of hadrons, nuclei and stars, and its full recovering at the limit of Schwartzchild's horizon **under the full preservation of quantum axioms merely subjected to a broader realization;** 6) We studied otherwise impossible advances in physics, chemistry and biology, including the exact representation of nuclear data, the achievement of an attractive force in valence electron bonds with ensuing exact representation of molecular data, and a new conception of life consisting of extended constituents under continuous EPR entanglement represented via hyperstructures [231]; 7) We showed the impossibility for the Copenhagen interpretation of quantum mechanics to solve our increasingly alarming environmental problems, such as recycling nuclear waste, achieving controlled nuclear fusions, and reaching the full combustion of fossil fuels; and pointed out their possible resolution under the EPR argument according to which "quantum mechanics is not a complete theory."

TABLE OF CONTENT

Fatti non foste a viver come bruti
ma per seguir virtute e canoscenza
Dante Alighieri

1. FOREWORD.

As it is well known, Nazism reached in 1935 the peak of its military, political as well as, lesser known, scientific power, the latter being due to the conception and construction of quantum mechanics by German scientists, such as M. Planck, I. Schrödinger and W. Heisenberg and others.

The R. M. Santilli Foundation (http://www.santilli-foundation.org) and the Family of Israel Foundation (http://www.i-b-r.org/translational-medicine.htm) organized, conducted and recorded the International Teleconference from September 1 to 5, 2020 (see Ref. [0]-[4] and proceedings papers [210]-[233]) for the study of old and new verifications of the historical view by Albert Einstein, expressed jointly with his graduate students B. Podolsky and N. Rosen, that *"quantum mechanics is not a complete theory,"* with ensuing expectation that a suitable completion of quantum mechanics would recover classical determinism at least under limit conditions (EPR argument) [1].

This Overview is intended to provide an outline of studies conducted over one century in old and new verifications of the EPR argument and their applications in physics, chemistry and biology, with particular reference to the search for possible resolutions of our increasing alarming environmental problems via the new sciences permitted by the EPR argument, which resolutions appear to be impossible via quantum mechanics (see representative references [2] - [235]).

In Section 2, we present a brief outline of the EPR argument and its historical objections, including Bell's inequalities; in Section 3, we review known insufficiencies of quantum mechanics in various fields; in Section 4, we review the experimental verifications of the validity of Bell's inequality for point-like particles under linear, local and potential interactions as well its inapplicability (rather than violation) for extended particles under non-linear, non-local and non-potential interactions; in Section 5, we review the first historical completions of quantum mechanics essentially along the EPR argument by W. Heisenberg, L. de Broglie, D. Bohm et al.; in Section 5, we outline the completion of quantum mechanics, chemistry and biology into the various fields; in Section 7, we review five different verifications of the EPR Argument by R. M. Santilli [210]-[214]; in Section 8 we review applications and predictions in physics, chemistry and biology, as well as the implications of the EPR argument for high energy scattering experiments; and in Section 9, we present what appears to be the ultimate implications of the EPR argument.

To properly present and document a century of research in the field, we have made an effort to provide free pdf downloads of:

i) Reprints of Santilli's 1998 and 2019 verifications of the EPR argument [210]-[214];

ii) Papers [6]-[188] by numerous authors that resulted in time to be significant for the proofs the EPR argument, including:

ii-1) Refs. [32]-[36] that are important for the axiomatic structure and elaboration of Lie-isotopic theories;

ii-2) Refs. [49]-[52] that contain original important contributions by various authors for the construction of hadronic mechanics;

ii-4) Refs. [53]-[67] that provide a step-by-step Lie-isotopic completion of conventional space-time symmetries, their implications for physical laws and the proof of their isomorphism with conventional symmetries, which references play an important role in the construction of explicit and concrete realizations of Bohm's hidden variables with ensuing verification of the EPR argument.

iii) Monographs by various authors [154] [201]-[207] that review the process leading to verifications [210]-[214] of the EPR argument;

iv) References in the first representation of nuclear magnetic moments, spin, stability and other data with the consequential prediction and initial verification of new clean nuclear energies (Sections 8.1 and 8.2);

v) Internet debates for anonymous expressions of personal views in cosmology [76], neutrino [91], gravitation [180], applications of the EPR argument in physics [208], and applications of the EPR argument in chemistry [209];

vi) Tutorial lectures in isomathematics [143], verifications of the EPR argument [144], Lie-admissible formulations [145], and hyperstructures [146]; and

vii) Proceedings of international meetings in the field [188]-[200], including: five *Workshops on on Lie-Admissible Formulations* conducted at Harvard University from 1978 to 1982; twenty five *Workshops on Hadronic Mechanics* conducted from 1983 on at various locations in the U.S.A., Europe and China; and three *International Conferences on non-Potential interactions and their Lie-admissible treatments,* the first conducted in 1981 at the Université d'Orleans, France, the second conducted in 1995 at the Castle Prince Pignatelli, Molise, Italy, and the third conducted in 2011 at Katmandu University, Nepal. Out of all these meetings, the author could identify only twelve proceedings available with links for free pdf download which illustrate the number of scientists who contributed in the construction of hadronic mechanics, chemistry and biology.

It should be stressed that, by no means, the content of this Overview is expected to be accepted by all participants of our 2020 Teleconference in the EPR Argument [0] because, particularly when dealing with fundamental open aspects, debates on qualified dissident views are essential for a serious scientific process.

2. THE EPR ARGUMENT.

A most mysterious experimental evidence in nature is the capability of particles to influence each other instantly at a distance. In view of the apparent influence in the 1930's by the Nazism, the scientific community of the time assumed that such an effect is predicted by quantum mechan-

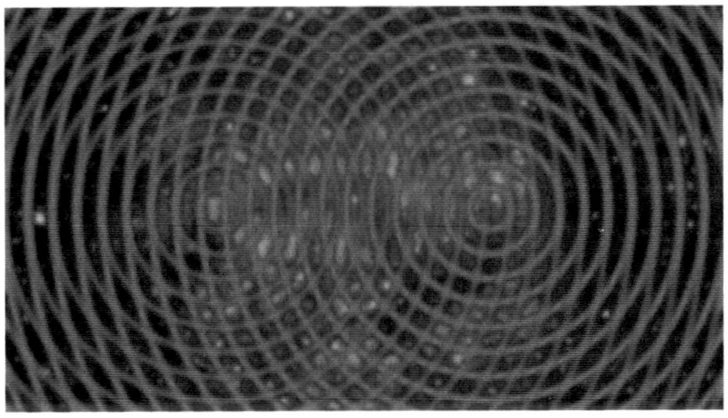

Figure 1: *A view of the new "EPR entanglement" introduced during the 2020 International Teleconference (see Section 7.2.3 for a technical treatment), which consists of particles under continuous and instantaneous communication via the overlapping of their wavepackets, thus without any need for superluminal communications[1], with ensuing contact, zero-range, non-linear, non-local and non-potential interactions represented via isomathematics [54] and isomechanics [24].*

ics, for which reason the effect continues to be called to this day *quantum entanglement.*

By contrast, Albert Einstein noted that the Schrödinger equation of quantum mechanics (for $\hbar = 1$)

$$[-\frac{1}{2m}\Delta_r + V(r)]\psi(r) = E\psi(r), \tag{1}$$

is strictly local, in the sense that it can only represent point-like particles at a distance in vacuum and, therefore, cannot predict the entanglement of particles, in which case the sole possible representation of the entanglement is that via superluminal communications i that would violate special relativity.

To avoid such a violation, Einstein, Podolsky and Rosen argued that "quantum mechanics is not a complete theory" [1].

The EPR argument was quickly criticized by N. Bohr [2] a few months following its appearance although, under a sufficiently deep scrutiny, Bohr did not truly address the EPR argument, namely, the possible existence of "elements of reality" in the universe beyond those of the atomic structure which would require a suitable completion of quantum mechanics.

Despite such a visible insufficiency, Bohr's opposition to the EPR argument was supported quite widely by numerous scientists apparently because of the Nazism scientific power of the time.

In 1964, J. S. Bell [3] proved an inequality which essentially established that *quantum mechanical systems of particles with spin 1/2 do not admit classical counterparts.* Bell's inequality was considered by the mainstream scientific community, with due exceptions, to be the final dismissal of Einstein's dream of recovering classical determinism, and set the current widely accepted view that quantum mechanics is valid for whatever conditions exist in the universe (see review [4] and its comprehensive literature).

3. CONCEPTUAL FOUNDATIONS

3.1. Insufficiencies of quantum mechanics in particle physics.

The Copenhagen interpretation of quantum mechanics can only represent particles as being point-like at a distance in vacuum, because Schrödinger's equation (1) is characterized by wavefunctions $\psi(r)$, potentials $V(r)$, Laplacians Δ_r and other quantities that can only be defined at a finite set of isolated points r. Consequently, contrary to a rather popular belief, the Copenhagen interpretation of quantum mechanics cannot consistently predict or represent particle entanglements without superliuminal communications, as correctly stated in the EPR argument [1].

The novel quantitative representation of particle entanglements studied at the 2020 teleconference [0] and called *EPR entanglement,* can be conceptually outlined as follows (see Figure 1 and Section 7.2.3 for a technical definition). Recall that the wavepacket of a particle fills up the entire universe with an intensity inversely proportional to the square of the distance. Hence, the entanglement of particles at a distance is characterized by the mutual penetration/overlapping of their wavepackets, under which conditions particles are in continuous instantaneous communication through the overlapping of their wavepackets without any need for superluminal speeds (Figure 1).

The technically challenging problem is that the interactions caused by the overlapping of the wavepackets of particles are: 1) Non-linear in the wavefunctions; 2) Non-local because defined in a volume; and 3) Not derivable from a potential because they are of contact, thus zero-range type, hereon referred to as *non-Hamiltonian interactions.*

Also recall that quantum mechanics is strictly Hamiltonian in the sense that interactions can only be represented via the Hamiltonian H. Consequently, the vast historical literature based on quantum mechanics (not listed in this Overview for brevity because easily identifiable with an internet search) represents particle entanglements with a Hamiltonian. By contrast, the 2020 Teleconference studied the alternative approach according to which the interactions occurring in particle entanglement should be represented with an operator *other* than the Hamiltonian, by therefore mandating a completion of quantum mechanics according to the EPR argument [1].

An important objective of the 2020 Teleconference has been that of reviewing half a century of research by R. M. Santilli as well as by various scholars in the completion of 20th century applied mathematics, from its sole validity at isolated points, into a covering mathematics providing a consistent representation of interactions 1), 2), 3) above (see the recorded Tutoring Lectures of Ref.[0]). Physical and chemical completions and applications of the new notion of particle entanglement were considered only thereafter.

Note that, from the basic axioms of the $SU(2)$-spin algebra and of Lie's theory at large, the validity of Bell's inequality [3] crucially depends for point-like particles under linear, local and potential interactions. Therefore, the study of systems of extended particles under non-linear, non-local and non-potential interactions automatically assures the *inapplicability* (rather than the violation of Bell's theorem), by therefore establishing rigorous grounds for the verification of the EPR argument.

A study of Gödel's incompleteness theorems presented at the 2020 Teleconference is available in Ref. [217] and in the recorded lecture by E. T. D. Boney [0].

The connection between causality and quantum mechanics was studied at the 2020 Teleconference by S. E. Johansen, see Ref. [224], and his recorded lecture [0].

The connection between the EPR argument and the minimum contradiction theory was studied at the 2020 Teleconference by A. A. Nassikas, see contributed paper [228] and his recorded lecture [0].

3.2. Insufficiencies of quantum mechanics in nuclear physics.

On serious scientific grounds, a theory can be claimed to be *exactly valid* for given systems ("elements of reality" [1]) if and only if the theory provides an exact representation of *all* experimental data of the systems considered from first axiomatic principles without the adulterations appearing in the contemporary physics literature via manipulated form factors, venturing of particles and/or entities not directly testable, and the like.

Under the above serious scientific conditions, quantum mechanics can indeed be considered to be exactly valid for the structure of the hydrogen atom. However, with the understanding that the *approximate* validity of quantum mechanics in nuclear physics is out of question, the assumption of quantum mechanics as being exactly valid for the nuclear

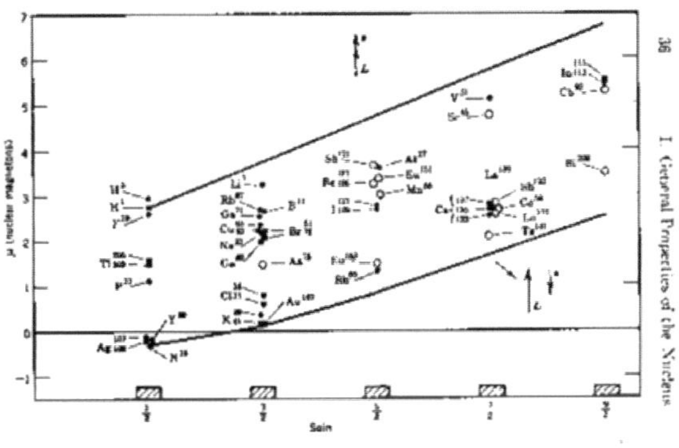

Figure 2: *A reproduction of the Schmidt limits providing a documentation of the deviations of the predictions of quantum mechanics from experimentally measured nuclear magnetic moments beginning with that of the smallest nucleus, the Deuteron. Similar insufficiencies exist for the representation of nuclear spins, nuclear stability and other data.*

structure implies the exiting from the boundaries of serious science due to the well known failure by quantum mechanics to achieve an exact representation of nuclear experimental data, such as nuclear magnetic moments, nuclear spins, nuclear stability (recall that the neutron is unstable.....), and other data, despite the use of billions of dollars of public funds in about one century of research. This insufficiency begins with the lack of representation of experimental data on the smallest nucleus, the Deuteron, and becomes embarrassing for large nuclei such as the Zirconium (Figure 2) [6].

An understanding of the origin of the indicated insufficiency can be reached via the comparison of atomic and nuclear structures. In the hydrogen atom, the proton and the electron are at such a large mutual distance to allow their effective point-like approximation, with the ensuing exact validity of the theory. By contrast, experimental data on nuclear

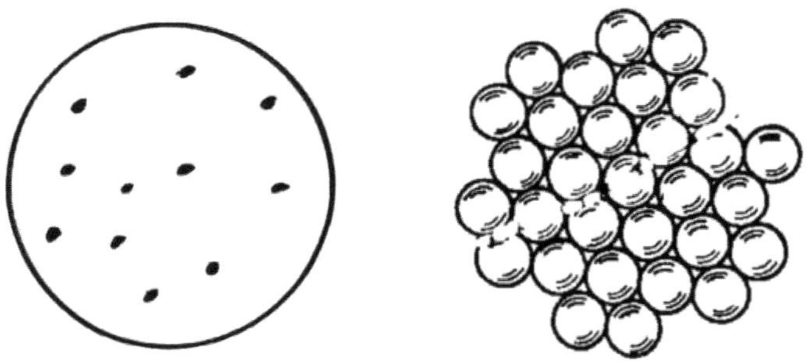

Figure 3: *An illustration on the left of the sole possible representation of nuclear structures by quantum mechanics due to its locality and linearity. An illustration on the right of the experimental reality establishing that nuclei are composed by extended and hyperdense nucleons in conditions of partial mutual overlapping with ensuing non-Hamiltonian interactions mandating a suitable completion of quantum mechanics [1].*

volumes and on the volume of individual nucleons, establishes that *nuclei are composed by a collection of extended and hyperdense nucleons in conditions of partial mutual penetration, thus EPR entanglement, of their charge distributions* (Figure 3) with ensuing non-Hamiltonian interactions (Section 3.1), with consequential need for a suitable completion of quantum mechanics.

3.3. Insufficiencies of quantum mechanics for irreversible processes.

Beginning with his Ph. D. studies at the University of Torino, Italy, in the mid 1960's, Santilli dedicated his research life to the proof of the EPR argument [1] because quantum mechanics cannot achieve a consistent representation of energy releasing processes, such as combustion, nuclear fusions and others. This is due to the fact that all energy releasing processes are irreversible over time, while quantum mechanics can only represent systems whose time reversal image does not violate causality (Figure 4, left view) due to the invariance under anti-Hermiticity of Heisenberg's equation for Hermitean operators A

$$i\frac{dA}{dt} = [A, H] = AH - HA = -[A, H]^\dagger. \qquad (2)$$

The century old objection against the verification of the EPR argument via irreversible processes, which is still widely accepted nowadays, is that irreversibility is 'illusory' (sic) because, when irreversible processes are reduced to their quantum mechanical elementary constituents, reversibility is fully regained. As part of his 1967 Ph. D. thesis [6]-[8], Santilli proved a number of theorems essentially stating that *a macroscopic irreversible system cannot be consistently decomposed into a finite number of quantum mechanical particles all in reversible conditions and, vice-versa, a finite number of quantum mechanical particles cannot recover a macroscopic irreversible system under the correspondence or other principles.*

An important discovery made during the 2020 Teleconference [0] is that *macroscopic irreversibility originates at the ultimate level of elementary particles, as established by the mere visual inspection of high energy scattering experiments at CERN, FERMILAB and other particle physics laboratories* (Figure 4, right view).

In view of the above insufficiency, Santilli proposed in 1967 (*loc. cit.*) the first known completion of quantum mechanics based on the embedding of Lie algebras of quantum mechanics with brackets $[A, B] = AB - BA$ into covering algebras with brackets (A, B) that are *Lie-admissible* (*Jordan-admissible*) according to the American mathematician A. A. Albert [8] when the attached anti-symmetric brackets $[A, B]^* = (A, B) - (B, A)$ verify the Lie algebra axioms (attached symmetric brackets $\{A, B\}^* = (A, B) + (B, A)$ verify the Jordan algebra axioms), with realization of the type

$$(A, B) = ARB - BSA = (ATB - BTA) + (AJB + BJA), \qquad (3)$$

Figure 4: *A view on the left of the irreversibility of combustion and its lack of representation via quantum mechanics due to its reversibility. A view on the right of the clear irreversibility of high energy particle collisions, with ensuing need for an irreversible completion of quantum mechanics (Section 8.7).*

where $R = T + J$ snd $S = -T + J$ are non-singular operators representing non-Hamiltonian interactions. The representation of irreversibility from first axiomatic principles is evidently assured by the violation of the invariance under anti-Hermiticity, $(A, B) \neq -(A, B)^\dagger$ which occurs whenever $R \neq S$. In particular, the Lie-isotopic operator T represents the non-Hamiltonian interactions of particle entanglement, while the Jordan-isotopic operator J represents the irreversibility of energy-releasing processes via a representation of the external terms in Lagrange's and Hamilton's equations that are completely absent in the Copenhagen interpretation of quantum mechanics (see Sections 6.2 and 7.6 for an overview).

3.4. Insufficiencies of quantum mechanics for antimatter.

Another majestic event in nature is given by the annihilation of particle-antiparticle pairs into light. The mechanism of this event cannot be rep-

Figure 5: *Another topic studied at the 2020 Teleconference [0] has been the inability by quantum mechanics to represent the mechanism of particle-antiparticle annihilation into light because charge conjugation characterizes antiparticles with the same positive energy of particles against Dirac's view [11] that antiparticles should have negative energy.*

resented via quantum mechanics because charge conjugation

$$\psi(r) \rightarrow \psi^c(r) = -\psi^\dagger(r), \tag{4}$$

characterizes antiparticles in the same Hilbert space of the original particles, as a result of which antiparticles have the same positive energy of particles. Consequently, a particle-antiparticle pair both having positive energy cannot possibly annihilate into light, besides being in conflict with P. A. M. Dirac historical hypothesis that antiparticles have *negative energy* [11].

Alternatively, there is no doubt that map (4) characterizes indeed a "charge conjugation." The question raised by Santilli is whether a pure "charge" conjugation can also be a consistent "particle-antiparticle" conjugation. As an example, there is no doubt that the charge conjugate of π^- meson is the positively charged meson π^+. However, in this case there is no known consistent representation of $\pi^- - \pi^+$ annihilation since they both have positive masses.

Similarly, the additional interpretation of map (4) as a particle-antiparticle map implies that the π^0 meson coincides with its antiparticle, but then what are the physical differences between massive neutral particles and their antiparticles?

These occurrences clearly suggest the need for an additional completion of quantum mechanics, this time, for a *particle-antiparticle conjugation* permitting a quantitative representation of particle-antiparticle annihilation that can be best achieved when all characteristics of antiparticles are opposite those of particles (see Section 7.7 for an overview).

3.5. Insufficiencies of quantum mechanics in chemistry.

With the full admission of the historical advances achieved by quantum chemistry in the past century, the advancement of basic scientific knowledge requires the indication of the basic insufficiency of quantum chemistry given by the absence of a quantitative model of valence bonds due to the inability to represent the *attractive force* between *identical* valence electrons (Figure 6), with consequential lack of a quantitative model of molecular structures that carries evident environmental implications [12].

In fact, the Schrödinger equation of valence electron pairs is given by

$$[-\frac{1}{m}\Delta_r + \frac{e^2}{r}]\psi(r) = E\psi(r), \tag{5}$$

where m is the reduced mass, thus solely allowing a repulsion caused by the equal electron charge $-e$ which is represented by $+e^2/r$. Simple calculations show that the *repulsive force* between two valence electrons at $10^{-13}cm = 10^{-15}fm$ mutual distance is given by

$$F = k\frac{e^2}{r^2} = (8.99 \times 10^9)\frac{(1.60 \times x10^{-19})^2}{(10^{-15})^2} = 230\ N, \tag{6}$$

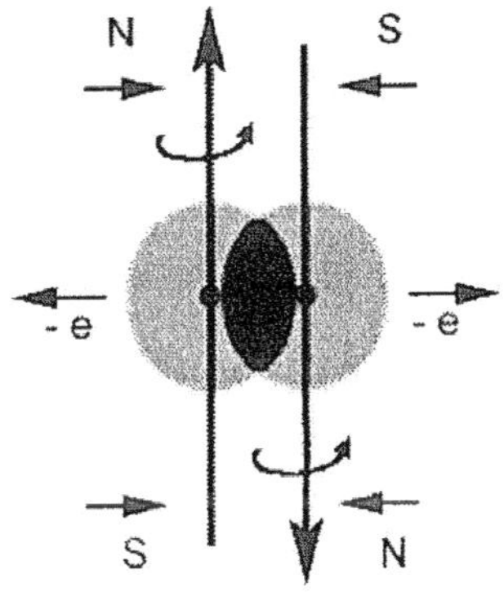

Figure 6: *A schematic view of the lack of a quantitative representation of molecular structures by quantum chemistry due to the large Coulomb repulsion between identical valence electrons according to Schrödinger equation (5) which is explicitly computed in equation (6).*

thus being so enormous for particle standards to prevent any realistic hope of being overcome by current models of valence bonds, thus confirming the need for a suitable completion of quantum chemistry according to the EPR argument [1].

The lack of a sufficiently strong attractive force between identical valence electrons has rather serious implications in chemistry, such as:

a) The lack of an exact representation of molecular binding energies with deviations of the order of 2% that, rather than being ignorable, corresponds to about $950\ kcal/mole$.

b) The prediction that all substances are ferromagnetic in view of the lack of strongly bonded valence electron pairs.

c) The inability to achieve a quantitative representation of molecular electric and magnetic moments, with consequential inability to identify the attractive force between molecules in their liquid state, and other insufficiencies [*loc. cit.*].

4. EXPERIMENTAL FOUNDATIONS.

An important lecture delivered at the 2020 Teleconference has been that by Gerald Eigen [13] (see also Ref. [235]) who presented various experiments that essentially provide a *clear experimental verification of Bell's inequality for point-like particles in vacuum, thus including electromagnetic interactions.*

Independently from the above tests, Matteo Fadel *et al* [14] conducted experiments similar to those of Ref. [13], but this time for the study of spin-correlations between space separated *atoms* of a Bose-Einstein condensate, by *measuring uncertainties below the bound of Heisenberg's uncertainty principle.*

Additionally, the ALICE experiments at CERN on *heavy ions* [15] have shown deviations from quantum mechanics bigger than those measured by experiments [14].

A view expressed at the 2020 teleconference is that tests [13]-[15] provide a conceptual and experimental verification of the EPR argument [1] because tests [13] confirm the acceptance by Einstein, Podolsky and Rosen of the validity of quantum mechanics for *point-like particles* in vacuum, while tests [14] [15] establish the existence of deviations from quantum mechanics for *extended particles.*

The main difference between tests [13] and [14] [15]] deal with the underlying interactions. In fact, for tests [13], the sole possible interactions are those at a distance, since the particles are dimensionless. By contrast, for the case of tests [14] [15], since the constituents are given by extended atoms and ions under the known compression caused by the Bose-Einstein condensate, the interactions are of the non-linear, nonlocal and non-potential type causing the *inapplicability* (rather than the violation) of Bell's inequality.

As we shall see, besides the above direct tests, there exist additional experimental evidence in support of the EPR argument in particle physics, nuclear physics, chemistry and other fields.

5. HISTORICAL EPR COMPLETIONS.

In general, the limitations of basic physical laws have been best identified by their originators. For example, following the discovery of the matrix realization of quantum mechanics on a Hilbert space $\mathcal{H}$ over the field of complex numbers $\mathcal{C}$, with historical time evolution (2), Werner K. Heisenberg identified the limitations of his own theory caused by its linearity, and studied what can be well called the *first non-linear completion of quantum mechanics* [16].

Quantum mechanics is a strictly Hamiltonian theory and, therefore, can only represent non-linear interactions with the Hamiltonian, resulting in eigenvalues equations of the type

$$H(r, p, \psi)\psi(r) = E\psi(r). \tag{7}$$

As part of his decades of studies on completions according to the EPR argument, Santilli pointed out at the 2020 Teleconference [0] that, despite its clear historical value, Heisenberg's non-linear completion does not allow the characterization of the constituents $\psi_k(r),\ k = 1, 2, ..., N$ of a bound state with non-linear internal interactions because non-linear equations (7) generally violate the superposition principle,

$$\psi(r) \neq \Pi_{k=1,2,...,N}\psi_k(r). \tag{8}$$

The lack of a superposition principle then implies serious limitations on the representation of experimental data, e.g., on how the neutron, which is naturally unstable, becomes stable when bonded to a proton in a stable nucleus.

Therefore, in order to preserve the superposition principle under completion, Santilli suggested that non-linear interactions should be represented with a suitably selected operator *other than the Hamiltonian.* In any case, as indicated earlier, non-linear interactions are of contact, zero-range type thus carrying no potential energy. Their representation with the Hamiltonian would then cause consistency problems in the axiomatic structure of the completed theory.

Despite the indicated limitation, Heisenberg's non-linear completion of quantum mechanics played an important role in Santilli's studies. In fact, there is a mention in the Comments of Teleconference [0] that Santilli had various mail contacts with Heisenberg in the early 1970's precisely on the superposition principle when Santilli was in the faculty of the Department of Physics of Boston University.

As it is well known, Louis V. P. R. de Broglie was one of the primary contributors to the development of the wave structure of particles in quantum mechanics. Jointly, de Broglie was one of the firsts scientists to identify the limitations of such a description caused by the locality of the wavefunction $\psi(r)$. The theory developed by de Broglie was semi-classical as a condition to by-pass Heisenberg's uncertainty principle and, as such, it can be considered to be one of the first attempts at recovering Einstein's determinism.

Unfortunately for science, de Broglie was dissuaded by mainstream scientists of the time to continue his studies on possible broadening of quantum mechanics. Nevertheless, de Broglie's ideas were resumed and further developed by David J. Bohm and the theory can be well called resulting the *de Broglie-Bohm non-local completion of quantum mechanics* [17].

This theory was discussed at the 2020 Teleconference [0] because of its important influence in the development of new completions of quantum mechanics. In particular, Santilli pointed out that, in his view, no consistent representation of the entanglement of particles can be achieved without a non-local theory since the overlapping of wavepackets occurs in a volume. Yet, to be effective, the representation of said non-locality should be done via an operator other than the Hamiltonian since said wave-overlapping carries no potential energy.

The third important historical completion of quantum mechanics studied at the 2020 Teleconference [0] is that by D. J. Bohm with his theory of *hidden variables* [18].

As it is well known, quantum mechanics is a probabilistic theory, with ensuing uncertainties in the position and momentum of particles. Bohm conjectured the possible existence of more fundamental physical laws hidden in the mathematics of quantum mechanics that would allow the theory to recover Einstein's determinism. Despite a large literature in the field, no concrete formulation of hidden variables, and related recovering of Einstein's determinism, was achieved up to the early 2000's . Nevertheless, Bohm's intuition on the lack of final character of the Copenhagen interpretation of quantum mechanics remained fundamental.

Santilli's view on Bohm's hidden variables presented at the 2020 Teleconference [0] is essentially the following:

1) Bell's inequality on the lack of existence of classical images should be considered valid for the infinite family of unitary equivalence of quantum mechanics. Consequently, the sole known possibility of bypassing Bell's inequality and achieving concrete realizations of hidden variables is the construction of a *non-unitary completion of quantum mechanics.*

2) Bell's inequality has been experimentally verified by G. Eigen and others [13] for electromagnetic interactions of point-like particles in vacuum. Consequently, the most promising applications of non-unitary completions of quantum mechanics are those for strong nuclear interactions for which quantum mechanics is known to be incomplete (Section 2).

3) As it is the case for Heisenberg's non-linear theory, and de Broglie-Bohm on-local theory, Bohm hidden variables should be realized via an *operator*— other than the Hamiltonian which is hidden in the axioms of quantum mechanics.

A geometric interpretation of hidden variables was presented at the 2020 Teleconference by O. A. Olkhov, see contribution [239] and his recorded lecture [0].

6. HADRONIC MECHANICS, CHEMISTRY AND BIOLOGY.

6.1. Hadronic mechanics.

It is evident that the completion of quantum mechanics according to Einstein, Podolsky and Rosen [1] has implications for all quantitative sciences, including physics, chemistry and biology. In order to initiate an expectedly long process, in his Ph. D. thesis at the University of Torino, Italy, in the mid 1960's, Santilli proposed the completion of quantum mechanics for irreversible processes via Lie-admissible completion of the quantum mechanical Lie algebras [6]-[8] according to A. A. Albert [9], and continued the studies at Harvard University under DOE support in the late 1970's with further developments of the Lie-admissible formulations as well as their Lie-isotopic particularization [19]-[22].

Thanks to important contributions by numerous mathematicians, physicists and chemists, the above studies achieved in the late 1990's a completion/covering of quantum mechanics known as *hadronic mechanics* [23]-[28] for a consistent representation of extended particles under non-linear, non-local and non-Hamiltonian interactions due to wave-overlapping, thus EPR entanglement.

An important feature of the studies herein considered is that *the Copenhagen interpretation of quantum mechanics resulted as being the simplest possible realization of quantum axioms, such as associativity, distributivity, linearity, locality, potentiality, etc., while hadronic mechanics is characterized by the progressive most general possible realizations of said axioms depending on the complexity of the considered "elements of reality"* [1], according to the following:

Classification of hadronic mechanics:

6-1-I) Quantum mechanics, with the familiar *Heisenberg's time evolu-*

tion (2) of an observable A in the infinitesimal form

$$i\frac{dA}{dt} = [A, H] = AH - HA = -[A, H]^\dagger, \quad A = A^\dagger, \quad H = H^\dagger, \tag{9}$$

and finite form

$$A(t) = UA(0)U^\dagger = e^{Hti}A(0)e^{-itH}, \quad UU^\dagger = U^\dagger U = I, \tag{10}$$

characterized by the Lie's theory with brackets $[A, H]$ for the representation of stable, thus time-reversal invariant systems of point-like particles in vacuum via the Hamiltonian H.

6-1-II) Isomechanics (see Refs. [23] [24], reviews in Section 2 of Ref. [25] and Tutoring Lecture I [143][144]), with the *iso-Heisenberg time evolution* in its infinitesimal form (first introduced in Eqs. (18), page 153, Ref. [22] and called *Heisenberg-Santilli isoequation*)

$$i\frac{dA}{dt} = [A\hat{,}H] = A\hat{T}H - H\hat{T}A = -[A\hat{,}H]^\dagger, \hat{T} > 0, \tag{11}$$

and finite form

$$A(t) = WA(0)W^\dagger = e^{H\hat{T}ti}A(0)e^{-it\hat{T}H}, \quad WW^\dagger \neq I, \tag{12}$$

characterized by the axiom-preserving Lie-Santilli isotheory with brackets $[A\hat{,}B] = A\hat{T}B - B\hat{T}A$, for the representation of stable (thus time-reversal invariant) hadrons, nuclei and stars composed by extended, therefore deformable constituents in conditions of deep mutual entanglement with conventional interactions represented with the Hamiltonian H, and non-linear, non-local and non-potential interactions represented with the isotopic element $\hat{T}$.

6.1-III) Genomechanics (see Refs. [23] [24], reviews in Section 3 of Ref. [25] and Tutoring Lecture IV [145]), with the *geno-Heisenberg equation,* also called *Santilli's Lie-admissible equation* in their infinitesimal form (first introduced in Eqs. (19), page 153, Ref. [22])

$$i\frac{dA}{dt} = (A, H) = A\hat{R}H - H\hat{S}A =$$

$$= (A\hat{T}H - H\hat{T}A) + (A\hat{J}H + H\hat{J}A) \neq -(A, B)^{\dagger},$$

$$\hat{R} = \hat{T} + \hat{J} \neq \hat{S} = -\hat{T} + \hat{J}, \ \ \hat{T} > 0, \ \ \hat{J} > 0, \tag{13}$$

and finite form

$$A(t) = WA(0)W^{\dagger} = e^{H\hat{R}ti}A(0)e^{-it\hat{S}H}, \ \ WW^{\dagger} \neq I, \tag{14}$$

characterized by the Lie-admissible/Jordan-admissible theory with product (3), for the representation of time-irreversible processes between extended constituents via: A)The conventional Hamiltonian H for the representation of conventional potential interactions; B) The Lie-isotopic operator $\hat{T}$ for the representation of non-Hamiltonian interactions due to mutual EPR entanglement as in Case II; and C) The Jordan-isotopic operator $\hat{J}$ for the representation of the external terms in the Lagrange's and Hamilton's equations that are completely missing in the Copenhagen interpretation of quantum axioms.

6.1-IV) Hypermechanics (see Ref. [27] [28] and Tutoring Lecture V [146]) characterized by the formulation of genomechanics in terms of T. Vougiouklis H_v hyperstructures [233] for the representation of irreversible systems with a large number of extended particles in deep EPR entanglement.

6.1-V) Isodual quantum, iso-, geno- and hyper-mechanics [29] for the representation of antiparticles in condition of increasing complexity via an anti-Hermitean map (called *isoduality* and denoted with the upper index d) of the preceding time evolutions, such as

$$i^d \times^d \frac{d^d A^d}{d^d t^d} = [A^d, H^d]^d = A^d \times^d H^d - H^d \times^d A^d, \tag{15}$$

providing a causal representation of antimatter with *negative energy* according to P. A. M. Dirac [11] because necessary to represent particle-antiparticle annihilation (see Section 3.4 and Figure 5, contribution [227] to the 2020 Teleconference and the recorded lecture by A. S. Muktibodh [0]).

6.2. Hadronic chemistry. Following the achievement of mathematical [23] and physical [24] maturity, as well as a number of experimental verifications [25] [26], hadronic mechanics was applied to chemistry, resulting in a completion-covering of hadronic chemistry for the representation of molecules composed by extended constituents in condition of EPR entanglement, called **hadronic chemistry** [30] which is also axiom-preserving, thus implying that *quantum and hadronic chemistry coincide at then abstract realization-free level.*

The most important function of hadronic chemistry is to show that the basic open problems in chemistry listed in Section 3.5 cannot be solved via quantum chemistry due to its linearity, locality and potentiality. By contrast, the achievement of an attractive force between identical valence electrons in molecular bonds has been possible [30] following the admission of the "element of reality" [1] given by the non-linearity, non-locality and non-potentiality of valence electron bonds (Figure 6). This new strong valence bond has permitted the achievement of exact representations of experimental data for the Hydrogen [31], water [32] and other molecules.

6.3. Hadronic biology. Studies in the field were initiated by C. R. Illert [176] who proved that a conventional three-dimensional Euclidean space $E(r, \delta, I)$, $r = (x, y, z)$, $\delta = I = Dag.(1, 1, 1)$ can indeed represent all possible *shapes* of seashells, but the *representation of their growth in time requires a three-dimensional two-valued space* here denoted $\hat{E}(\hat{r}, \delta, I)$, $\hat{r} = (\hat{r}_k)$, $\hat{r}_k = (r_k^1, r_k^2), k = 1, 2, 3.$

In order to initiate the expectedly long process to represent the extreme complexity of the DNA code, and by recalling that all living organisms are irreversible over time, thus requiring genomathematics, Santilli introduced the representation of biological structures via three-dimensional multi-valued geno-Euclidean genospaces $\hat{E}(\hat{r}, \hat{\delta}, \hat{I})$ over genofields with genounits having an arbitrary number N of ordered elements, $\hat{I} = Diag.(I_1, I_2, ..., I_N)$ [34], with classical operations, e.g., the genoproduct of two observables $A > B$ yields N ordered results [27].

T. Vougiouklis formulated the preceding results via the reformulation of genomathematics in terms of his H_v-hyperstructures on a hyperfield formulated in terms of hyperoperations called *hopes* (see Tutoring Lecture V [146] and Ref. [233]).

Vougiouklis' hyperformulation allowed the introduction at the 2020 Teleconference [0] of a new conception of living organisms presented in Ref. [231] as being composed by a very large number of molecules all in EPR entanglement, thus being in continuous and instantaneous communication, resulting in such a complex structure that can only be quantitatively represented with the most complex mathematics developed by the human mind.

It is significative that the 2020 Teleconference [0] ended up with the view: *There is hope that 'hope' can represent life.*

7. VERIFICATIONS OF THE EPR ARGUMENT.

7.1. Foreword.

A main topic of the 2020 Teleconference [0] has been the study of *isotopic verifications* [210]-[214] of the EPR argument [1], namely, verifications based on the preservation of the basic axioms of quantum mechanics and their most general possible realization which verifications are the outcome of the life-long research by R. M. Santilli with contributions by numerous scholars.

In this section, we review the indicated isotopic verifications in a lan-

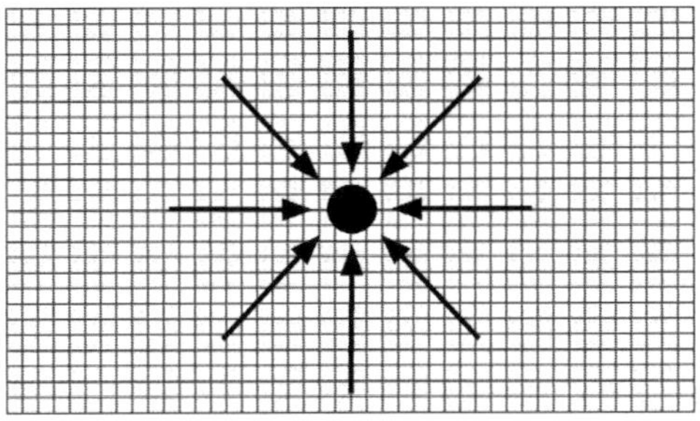

Figure 7: *This figure provides a conceptual rendering of the central notion used for the verifications of the EPR argument, namely, the inward 'pressure' experienced by 'extended' particles in 'all radial directions' when immersed in hyperdense media, such as a proton in the core of a star, which pressure evidently restricts uncertainties in favor of Einstein's determinism first identified in the 1981 paper [47] and then studied in detail in Refs.[210]-[214].*

guage as simple as possible, including a rudimentary review of the needed basic formalism for minimal self-sufficiency of this Overview, as well as for the specialization of the formalism to the EPR problem.

A central concept of this section is that *extended* particles immersed within a hyperdense medium, as it is the case for a proton in the core of a star, experience a *pressure in all radial directions* proportional to the density of the medium (Figure 7) which evidently restricts Heisenberg's uncertainties in favor of Einstein's determinism [1]. Note that such a notion of pressure is completely absent in the 20th century physics due to the point-like approximation of particles.

With the understanding that the participants to the 2020 Teleconference are available for consultations, including this author, it should be indicated that a full understanding of the isotopic verifications reviewed in this section requires a technical knowledge of hadronic mechanics [24].

7.2. Recovering of classical images.

7.2.1. Rudiments of isomathematics. As recalled earlier, Bell [3] proved the lack of existence of classical images for systems of quantum mechanical point-like particles with spin $1/2$.

Since the systems studied by Bell are described by the time-reversible Heisenberg law (9)-(10) (Sections 3 and 6), Santilli studied in the 1998 paper [210] systems characterized by isomathematics (Tutoring Lecture I [143]) and related isomechanics [23] [24] with iso-Heisenberg's time evolution (11)-(12) for which the represented systems are equally time-reversible, yet they admit internal non-linear, non-local and non-potential interactions represented with the isotopic element $\hat{T}$.

Additionally, Bell [3] used the conventional $SU(2)$-spin symmetry for the derivation of his result. Consequently, Santilli constructed the Lie-isotopic completion of the various branches of Lie's theory, including the isotopies of universal enveloping associative algebras, Lie algebras, and Lie groups [22], resulting in a theory nowadays known as the *Lie-Santilli isotheory* [21]-[25] [50] [55] [201] (see also contribution [226] to the 2020 Teleconference and the recorded lecture by A. S. Muktibodh [0]).

The foundation of the Lie-Santilli isotheory is given by the *universal isoenveloping isoassociative isoalgebra* with *isoproduct* between arbitrary quantities A, B (first introduced in Eq.(5), page 71, Ref. [22])

$$A\hat{\times}B = A\hat{T}B, \hat{T} > 0, \tag{16}$$

where $\hat{T}$, called the *isotopic element,* is solely restricted by the condition of being positive-definite, but can otherwise possess an arbitrary (non-singular) dependence on all needed local variables of interior dynamical problems, such as a dependence on time t, coordinates r, momentum p, energy E, density μ, temperature τ, pressure γ, wavefunctions ψ, their derivatives $\partial_r\psi$, etc. $\hat{T} = \hat{T}(t, r, p, E, \mu, \tau, \gamma, \psi, \partial\psi, ...)$, which dependence is hereon omitted for brevity.

It is evident that product (16) verifies the axioms of associativity and distributivity, although in the following more general form

$$(A\hat{\times}B)\hat{\times}C = A\hat{\times}(B\hat{\times}C),$$
$$A\hat{\times}(B + C) = A\hat{\times}B + A\hat{\times}C, \tag{17}$$

for which reason it is called the isoproduct.

It should be indicated upfront that *the primary function of isoproduct (16) is that of providing novel explicit and concrete realizations of Bohm's hidden variables [18] via the the isotopic element $\hat{T}$ which is "hidden" in the associative and distributive laws as clearly indicated by Eqs. (17).*

Santilli realized in the early 1978 that, despite its simplicity, isoproduct (16) requires for consistency the isotopic completions of *all* aspects of 20th century applied mathematics with no known exception. As an illustration of the Lie-Santilli isoalgebra, we indicate the isotopic completion $\hat{L}$ of an N-dimensional Lie algebra L with Hermitean generators X_k, $k = 1, 2, ..., N$, characterized by the *isocommutator rules*

$$[\hat{X}_i \hat{,} \hat{X}_j] = \hat{X}_i \hat{\times} \hat{X}_j - \hat{X}_j \hat{\times} \hat{X}_i =$$
$$= \hat{C}_{ij}^k (t, r, p, E, \mu, \tau, \psi, \partial\psi, ...) \hat{\times} \hat{X}_k, \tag{18}$$

whose verification of Lie algebra axioms is evident. Nowadays, Lie-Santilli isoalgebras are classified into *regular* when the structure quantities $\hat{C}$ are constant, and *irregular* when they are function of the local variables [55].

It should be noted, as illustrated below, that regular Lie-Santilli isoalgebras can be obtained via non-unitary transformations of the original Lie algebras, while irregular isoalgebras cannot, thus constituting new realizations of Lie's axioms.

The elaboration of isocommutation rules (18) with the conventional mathematics used for Lie's theory soon proved to lead to serious inconsistencies. This occurrence mandated the construction of isomathematics with the following main features:

1) The multiplicative unit of quantum mechanics $\hbar = 1$ is no longer invariant for isomechanics due to its non-unitary structure, Eq. (12). The consistent multiplicative unit under isoproduct (16) is given by the *isounit*

$$\hat{I} = 1/\hat{T}, > 0 \;\; \hat{I}\hat{\times}A = A\hat{\times}\hat{I} = A; \tag{19}$$

2) The isotopy of the unit $1 \rightarrow \hat{I}$ mandates, for consistency, the corresponding isotopic completion of numeric fields $F(n, \times, I)$ into *isofields* $\hat{F}(\hat{n}, \hat{\times}, \hat{I})$, first introduced in Ref. [33] of 1993, with *isonumbers* $\hat{n} = n\hat{I}$ where $n \in F$ equipped with isoproduct (16) and isounit (19);

3) The elaboration of the Lie-Santilli isotheory needs to be done, also for consistency, via the *isofunctional isoanalysis* initiated in Refs. [47] [48] of 1992 (see Refs. [23] [24] for a general treatment) with expressions, for instance, for the *isoexponent*

$$\hat{e}^X = (e^{X\hat{T}})\hat{I} = \hat{I}(e^{\hat{T}X}); \tag{20}$$

4) Recall that non-unitary transformations on conventional spaces over conventional fields generally violate causality[24] [212]. To regain causality, all non-unitary transformations can be reformulated into the *isounitary isotransforms*

$$U = \hat{U}\hat{T}^{1/2}, \;\; UU^\dagger \neq I \;\rightarrow\; \hat{U}\hat{\times}\hat{U}^\dagger = \hat{U}^\dagger\hat{\times}\hat{U} = \hat{I}, \tag{21}$$

for which the non-unitary time evolution(12) is identically reformulated in the correct isounitary form

$$A(t) = \hat{W}\hat{\times}A(0)\hat{\times}\hat{W}^\dagger = \hat{e}^{Hti}\hat{\times}A(0)\hat{\times}\hat{e}^{-itH},$$

$$\hat{W}\hat{\times}\hat{W}^\dagger = \hat{W}^\dagger\hat{\times}\hat{W} = \hat{I}; \tag{22}$$

5) For consistency, scalar quantities must be elements of isofields, with ensuing expression for *isocoordinates* $\hat{r} = r\hat{I}$ and generic expression for *isofunction*

$$\hat{f}(\hat{r}) = [f(r\hat{I})]\hat{I}; \tag{23}$$

6) Also for consistency, Lie-Santilli isoalgebras must be formulated on *isospaces over isofields,* (first formulated in Ref. [34] with reference, as an illustration, for non-relativistic formulations to the *iso-Euclidean isospace* $\hat{E}(\hat{r}, \hat{\delta}, \hat{I})$ with *isocoordinates* $\hat{r} = r\hat{I}$, $r = (x, y, z)$, *isometric* $\hat{\delta} = (\hat{T}\delta)\hat{I}$ where $\delta = Diag.(1, 1, 1)$ and *isoinvariant* as an element of the isofield of isoreal isonumbers $\hat{\mathcal{R}}$

$$\hat{r}^{\hat{2}} = \hat{r}^i \hat{\times} \hat{\delta}_{i,j} \hat{\times} \hat{r}^j = (\frac{x^2}{n_1^2} + \frac{y^2}{n_2^2} + \frac{z^2}{n_3^2})\hat{I}, \tag{24}$$

and the *iso-Minkowski isospace* for relativistic treatments $\hat{M}(\hat{x}, \hat{\eta}, \hat{I})$, with *space-time isocoordinates* $\hat{x} = x\hat{I} = (x, y, z, t)\hat{I}$, *space-time isometric* $\hat{\eta} = \hat{T}\eta$, $\eta = Diag.(1, 1, 1, -1)$ and *space-time isoinvariant*

$$\hat{x}^{\hat{2}} = \hat{x}^{\mu} \hat{\times} \hat{\eta}_{\mu\nu} \hat{\times} \hat{x}^{\nu} = (\frac{x^2}{n_1^2} + \frac{y^2}{n_2^2} + \frac{z^2}{n_3^2} - t^2\frac{c^2}{n_4^2})\hat{I}; \tag{25}$$

7) Since the multiplicative isounit is generally dependent on local isocoordinates, $\hat{I} = \hat{I}(\hat{r}, ...)$, the elaboration of Lie-Santilli isoalgebras via the conventional Newton-Leibnitz calculus leads to axiomatic inconsistencies (see also Section 7.3), and must be lifted into the *isodifferential isocalculus,* first introduced by Santilli in Ref. [34] (see also monographs. [207], contributions [222] [223] to the 2020 Teleconference, and the recorded lectures by S. Georgiev [0]), with basic *isodifferential*

$$\hat{d}\hat{r} = \hat{T}d[r\hat{I}(r, ...)] = dr + r\hat{T}d\hat{I}' \tag{26}$$

and *isoderivatives*

$$\frac{\hat{\partial}\hat{f}(\hat{r})}{\hat{\partial}\hat{r}} = \hat{I}\frac{\partial\hat{f}(\hat{r})}{\partial\hat{r}}; \tag{27}$$

8) Contrary to a popular belief that isomathematics is excessively complex for physicists, *all aspects of (regular) isomathematics can be very easily*

constructed via a non-unitary transform of the corresponding aspect of 20th century applied mathematics. For instance, the trivial unit 1 can be transformed into the isounit $\hat{I}$, the conventional associative product AB can be transformed into the isoproduct $A\hat{T}B$, etc. [35]

$$1 \rightarrow U1U^{\dagger} = \hat{I} = 1/\hat{T} \neq I, \ \hat{T} = (UU^{\dagger})^{-1},$$

$$AB \rightarrow U(AB)U^{\dagger} = (UAU^{\dagger})(UU^{\dagger})^{-1}(UBU^{\dagger}) = A'\hat{T}B',$$

$$E^X \rightarrow \hat{e}^X = (e^{X\hat{T}})\hat{I} = \hat{I}(e^{\hat{T}X}), etc.; \tag{28}$$

9) The crucial invariance of the numeric values of the isounit and isotopic element representing non-Hamiltonian interactions is verified by the appropriate time evolution, the isounitary form (21)-(22) [35]

$$\hat{I} \rightarrow \hat{U}\hat{\times}\hat{I}\hat{\times}\hat{U}^{\dagger} = \hat{I}' \equiv \hat{I},$$

$$A\hat{\times}B = A\hat{T}B \rightarrow \hat{U}\hat{\times}(A\hat{\times}B)\hat{\times}\hat{U}^{\dagger} = A'\hat{\times}'B' = A'\hat{T}'B', \ \hat{T}' \equiv \hat{T}. \tag{29}$$

Following the isotopic completion of Lie's theory in the period 1978-1982 [19]-[22], the step-by-step isotopies of conventional space-time mathematics, Santilli constructed in Refs. [60]-[70] the isosymmetries leaving invariant isospacetime (26).

These studies established that, rather then being violated for line element (25), the Lorentz symmetry remains valid because its isotopic image $\hat{SO}(3.1)$ is isomorphic to the original symmetry $SO(3.1)$ [60], and the same holds for all continuous as well as discrete space-time symmetries [23] [24].

We here merely recall for future comments: the *Lorentz-Santilli isotransformations* introduced in the 1983 paper [60], here indicated in the $(3, 4)$-plane (see Ref. [24] for their full formulation)

$$x^{1'} = x^1, \ x^{2'} = x^2,$$

$$x^{3'} = \hat{\gamma}(x^3 - \hat{\beta}\frac{n_3}{n_4}x^4), \quad x^{4'} = \hat{\gamma}(x^4 - \hat{\beta}\frac{n_4}{n_3}x^3), \tag{30}$$

where

$$\hat{\beta} = \frac{v_3/n_3}{c/n_4}, \quad \hat{\gamma} = \frac{1}{\sqrt{1-\hat{\beta}^2}}; \tag{31}$$

the second order iso-Casimir invariant of the *Lorentz-Poincaré-Santilli isosymmetry* [66]-[69]

$$\hat{C}_2 = (\hat{\eta}^{\mu\nu} P_\mu P_\nu)\hat{I}^c =$$

$$(n_1^2 P_1^2 + n_2^2 P_2^2 + n_3^2 P_3^2 - n_4^2 P_4^2)\hat{I}^c = m^2 C^2, \tag{32}$$

where, from line element (25), C s the local speed of light left invariant by isosymmetry (30),

$$C = \frac{c}{n_4}, \tag{33}$$

and $\hat{I}^c$ is now the *contravariant isounit,* thus being equal to $\hat{T}$,

$$\hat{I}^c = \hat{T} = 1/\hat{T}^c = Diag.(\frac{1}{n_1^2}, \frac{1}{n_2^2}, \frac{1}{n_3^2}\frac{1}{n_4^2}); \tag{34}$$

and the *iso-Klein-Gordon isoequation* [24] derivable from iso-Casimir invariant (32) which, for the simplest possible assumptions $n_k = 1$, $k = 1, 2, 3$, $\hbar = 1$ and m_4 is a positive quantity given by

$$(-\frac{1}{c}\frac{\partial^2}{\partial t^2} + \frac{1}{n_4^2}\bigtriangledown)\hat{T}^c\hat{\psi}(\hat{x}) =$$

$$= (\frac{m^2}{n_4^2}\frac{c^2}{n_4^2})\hat{T}^c\hat{\psi}(\hat{x}) = (\bar{m}^2 C^2)\hat{T}^c\hat{\psi}(\hat{x}), \tag{35}$$

where

$$\bar{m} = \frac{m}{n_4^2} \tag{36}$$

is the *isorenorrmalized energy* of an extended particle within a physical medium with $n_1^2 = n_2^2 = n_3^2 = 1$ holding for the particular case in which the particle is perfectly spherical and the medium is homogeneous and isotropic [24] (see the formal treatment in Section 8.4.4, Eq. (133) of Isoaxiom IV).

Finally, the isotopies $\hat{SU}(2)$ of the $SU(2)$-spin symmetry were constructed in Refs. [64] [65] (see the review in Section 3, Ref.[213]).

7.2.2. Recovery of classical images. Thanks to the above advances, Santilli introduced in paper [210] the following simplest possible realization of Bohm's hidden variables

$$\hat{U}U^\dagger = Diag(\lambda, 1/\lambda) \neq I, \quad Det\hat{I} = 1, \quad \lambda > 0, \tag{37}$$

applied to Pauli's matrices

$$\sigma_k \rightarrow \hat{\sigma}_k = U\sigma_k U^\dagger, \tag{38}$$

with explicit form called *Pauli-Santilli isomatrices*

$$\hat{\sigma}_1 = \begin{pmatrix} 0 & \lambda^{-1} \\ \lambda & 0 \end{pmatrix}, \quad \hat{\sigma}_2 = \begin{pmatrix} 0 & -i\lambda^{-1} \\ i\lambda & 0 \end{pmatrix}, \quad \hat{\sigma}_3 = \begin{pmatrix} \lambda & 0 \\ 0 & -\lambda^{-1} \end{pmatrix}, \tag{39}$$

with regular isoalgebra

$$[\hat{\sigma}_i \hat{,} \hat{\sigma}_j] = \hat{\sigma}_i \hat{T} \hat{\sigma}_j - \hat{\sigma}_j \hat{T} \hat{\sigma}_i = i2\epsilon_{ijk}\hat{\sigma}_k, \tag{40}$$

establishing the isomorphism $\hat{SU}(2) \approx SU(2)$, and *isoeigenvalue isoequations* (where all products, thus including squares, are isotopic)

$$\hat{\sigma}_3 \hat{\times} |\hat{b}> = \pm|\hat{b}>, \tag{41}$$

$$\hat{\sigma}^{\hat{2}} = (\hat{\sigma}_1 \hat{T} \hat{\sigma}_1 + \hat{\sigma}_2 \hat{T} \hat{\sigma}_2 + \hat{\sigma}_3 \hat{T} \hat{\sigma}_3) \hat{T} |\hat{b}>= 3|\hat{b}. \tag{42}$$

thus confirming the spin $1/2$ for *isoparticles,* namely, extended particles in deep EPR entanglement which are characterized by isounitary isoirreducible isorepresentation of the Lorentz-Poincaré-Santilli isosymmetry $\hat{P}(3.1)$ [66]-[69] (see Section 2 of paper [213] for a review).

Bell's proof of the lack of existence of classical images for two particles with spin $1/2$ is reduced a quantum mechanical (qm) quantity D^{qm} which, when computed via the use of Pauli's matrices, verifies the expression

$$D^{qm} \leq 2. \tag{43}$$

Santilli conducted in paper [210] a step-by-step *non-unitary/isounitary,* transformation of Bell's derivation along rules (28) for two isoparticles described by hadronic mechanics (hm) with Bohm's hidden variables λ_1, λ_2 in realization (37), resulting in the following property

$$D^{hm} = \frac{1}{2}(\frac{\lambda_1}{\lambda_2} + \frac{\lambda_2}{\lambda_1}) D^{qm}. \tag{44}$$

Since the factor $\frac{1}{2}(\lambda_1/\lambda_2 + \lambda_2/\lambda_1)$ can assume values arbitrarily larger than 2, Santilli proved in Ref. [210] that *a systems of isoparticles with spin 1/2 in condition of mutual penetration, thus EPR entanglement and ensuing non-Hamiltonian interactions does admit classical images* with specific examples identified in paper [210].

The invariance of the results under the the time evolution of the theory was also proved according to rules (39).

7.2.3. Realization of the EPR entanglement. It should be noted that realization (28) of hidden variables is indeed the most elementary possible because the notion of EPR entanglement (Figure 1) is characterized by the isotopic element $\hat{T}$ with realization for the hadronic bound state of two isoparticles (such as the Deuteron) of the type

$$\hat{T} = \Pi_{k=1,2} Diag.(1/n_{1k}^2, 1/n_{2k}^2, 1/n_{3k}^2, 1/n_{4k}^2) \times$$

$$\times exp[-\Gamma(\psi, \hat{\psi}, ...)] > 0, \tag{45}$$

where $n_{k\alpha}^2$, , $\alpha = 1, 2, 3, k = 1, 2$, where: $n_{\alpha k}^2 = 1$ represents the deformable semi-axes of the extended k-particle normalized to the values for the sphere; n_{4k}^2 represents the density of the k-particle normalized to the value $n_{4k}^2 = 1$ for the vacuum; and the exponent is a positive-definite function representing non-linear, non-local and non-potential interactions with realizations of the type

$$exp[-\Gamma(\psi, \hat{\psi}, ...)] = exp[-|\psi/\hat{\psi}| \int \hat{\psi}_1 \hat{\psi}_2^\dagger d^3r,] \tag{46}$$

where ψ is a quantum mechanical wave function, and $\hat{\psi}$ is the completed wavefunction under isotopy.

In conclusion, Ref. [210] has shown that the historical intuition of *hidden variable* by David Bohm [18] is correct and truly fundamental because it characterizes the completion of quantum into hadronic mechanics resulting in a basically new notion of particle entanglement with momentous applications indicated in the next sections.

It should be indicated that realization (45) of the isotopic element constitutes the first known *explicit representation of strong interactions*, in view of its capability of representing all data for the Deuteron and other nuclei (Section 8.2.5).

More generally, *the isotopic element constitutes an explicit and concrete realization of the 'fifth interactions'* here referred to the infinite family of interactions that are not representable with a Hamiltonian.

Some cosmological implication of the EPR argument originating from the classification of elementary particles via diagrams of the geometric SU(3) algebra was presented at the 2020 Teleconference [0] by E. Trell (eee contributed paper [232]).

7.3. Recovering of Einstein's determinism.

7.3.1. Rudiments of isomechanics. The proof of the recovering of Einstein's determinism achieved in the 2019 paper [211] requires the use of

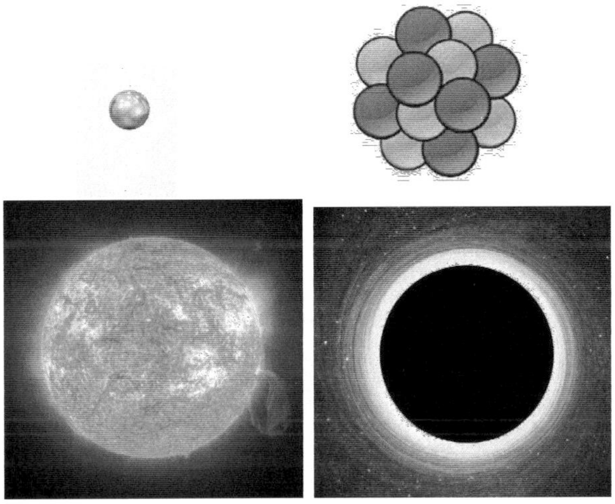

Figure 8: *Experimental evidence establishes that the wavepacket of particles has approximately the same size* 10^{-13} *cm as that of all hadrons (top left). Consequently, hadrons are composed by wavepackets in deep EPR entanglement (Sections 3.1 and 7.2.3) called 'hadronic medium.' A corresponding structure holds for nuclei, stars and black holes due to partial or deep mutual penetration of hadron constituents. The realization via the isotopic element of hadronic mechanics of the pressure exercised by the hadronic medium on individual constituents (Figure 7) implies the progressive weakening of Heisenberg's uncertainties first identified in Ref. [47] of 1981, with progressive recovering of Einstein's determinism [1] in the structure of hadrons, nuclei and stars and its full recovering at the limit of gravitational collapse studied in detail in Refs. [210]-[214]. Note that Heisenberg's uncertainties remain valid for the center of mass of hadrons, but not so for black-holes, thus illustrating the indicated full recovering of Einstein's determinism.*

isomathematics outlined in the preceding section plus the use of isomechanics [24] whose basic elements can be briefly outlined for the non-initiated reader as follows:

1) The basic unit of isomechanics is given by the completion of Planck's unit $\hbar = 1$ into the isounit $\hat{I}$ which is the inverse of the isotopic element (45)

$$\hat{I} = 1/\hat{T} = \Pi_{k=1,2}Diag.(n_{1k}^2, n_{2k}^2, n_{3k}^2, n_{4k}^2) > 0, \qquad (47)$$

where exponent (46) is incorporated into the n-characteristic quantities. Isounit (47) represents the *volume* and *density* of isoparticles as well as the impossibility for *extended* protons in the core of a star to solely have discrete energy exchanges due to the extreme local densities and pressures, in favor of integro-differential energy exchanges;

2) The Lie-Santilli isotheory has be formulated on a *Hilbert-Myung-Santilli isospace* [45] $\hat{\mathcal{H}}$ over the isofield of isocomplex isonumbers $\hat{\mathcal{C}}$ [33];

3) Isostates $\hat{\psi}(\hat{r}) \in \hat{\mathcal{H}}$ should have *isonormalization*

$$< \hat{\psi}(\hat{r})|\hat{\times}|\hat{\psi}(\hat{r}) >=< \hat{\psi}(\hat{r})|\hat{T}|\hat{\psi}(\hat{r}) >=$$

$$= \int_{-\infty}^{+\infty} \hat{\psi}^{\dagger}(\hat{r})\hat{T}\hat{\psi}(\hat{r})d\hat{r} = \hat{T}, \qquad (48)$$

were one should note the need for isointegrals and isodifferentials (23);

4) The *isoexpectation values* of an observable A are given by

$$\hat{<}A\hat{>} =< \hat{\psi}(\hat{r})|\hat{\times}A\hat{\times}|\hat{\psi}(\hat{r}) >=< \hat{\psi}(\hat{r})|\hat{T}A\hat{T}|\hat{\psi}(\hat{r}) >; \qquad (49)$$

5) The realization of the linear momentum on isospaces over isofield via the conventional differential calculus leads to major inconsistencies. The correct formulation of the *isolinear isomomentum* is that via isoderivative (27) and it is given by

$$\hat{p}\hat{\times}\hat{\psi}(\hat{r}) = -\hat{i}\hat{\times}\hat{\partial}_{\hat{r}}\hat{\psi}(\hat{r}) = -i\hat{I}\partial_{\hat{r}}\hat{\psi}(\hat{r}), \qquad (50)$$

thus providing a forceful illustration of the need for the novel isodifferential calculus [34];

6) The *isocanonical isocommutation rules* are consequently given by

$$[r^i \hat{,} p_j] = \hat{i} \hat{\times} \hat{\delta}^i_j = i \hat{I} \delta^i_j, \quad [r^i \hat{,} r^j] = [p_i \hat{,} p_j] = 0, \tag{51}$$

by illustrating that isomechanics is *isolinear*, that is, linear on isospace over isofields because $[p_i \hat{,} p_j] = 0$, although the theory is highly non-linear when *projected* into our conventional spaces over conventional fields because, in that case, $[p_i, p_j] \neq 0$;

7) The *iso-Schrödinger isoequation* (also called *Schrödinger-Santilli isoequation*) is then given by [20]-[24]

$$\hat{H} \hat{\times} \hat{\psi}(\hat{r}) = [-\frac{1}{2m} \hat{\Delta}_{\hat{r}} + \hat{V}(\hat{r})] \hat{\times} \hat{\psi}(\hat{r}) = \hat{E} \hat{\times} \hat{\psi}(\hat{r}) = E \hat{\psi}(\hat{r}), \tag{52}$$

and now represents extended, deformable and hyperdense particles at all levels, including isocoordinates $\hat{r}$, isostates $\hat{\psi}(\hat{r})$, isotopic element $\hat{T}$, isopotentials $\hat{V}(\hat{r})$ iso-Laplacian $\hat{\Delta}_{\hat{r}}$, etc. In particular, isostates verify the isosuperposition principle on isospaces over isofields

$$\hat{\psi}(\hat{r}) = \Sigma_{k=1,2} \hat{\psi}_k(\hat{r}), \tag{53}$$

because the theory is isolinear as indicated above. Therefore, hadronic mechanics resolves the inability by Heisenberg's non-linear completion of quantum mechanics [16] to represent the constituents of a bound state of particles with non-linear internal forces discussed in Section 5.

7.3.2. Confirmation of the EPR argument. It is evident that the recovering of classical images in paper [210] established the foundation for the recovering of Einstein's determinism. These studies were initiated by Santilli in the 1981 paper [47], continued in various papers due to the need to achieve maturity in the formulation of isomathematics (see monographs [23] [24]), and concluded in the 2019 paper [211] via the following simple isotopy of the conventional derivation of Heisenberg's uncertainties

$$\Delta r \Delta p = \frac{1}{2} | < \hat{\psi}(\hat{r}) | \hat{\times} \, [\hat{r} \hat{,} \hat{p}] \, \hat{\times} | \hat{\psi}(\hat{r}) > \approx \frac{1}{2} \hat{T} \ll 1, \tag{54}$$

where the property $T \ll 1$ is established by all fits of experimental data to date [25] [26] (see Section 7.6 for the analytic aspects).

By recalling that the value of the isotopic element is inversely proportional to the increase of the density [211], isodeterministic principle (47) establishes *the progressive validity of Einstein's determinism in the interior of hadrons, nuclei and stars, and its full achievement in the interior of gravitational collapse* (Figure 8).

The latter result is due to the fact that the isotopic element admits a realization in terms of the space component of Schwartzchild's metric [211]

$$\hat{T} = \frac{1}{1 - \frac{2M}{r}} = \frac{r}{r - 2M}, \tag{55}$$

where M is the gravitational mass of the body considered with ensuing *isodeterministic isoprinciple*

$$\Delta \hat{r} \Delta \hat{p} \approx \hat{T} = \frac{r}{r - 2M} \Rightarrow_{r \to 0} = 0. \tag{56}$$

Note that the center-of-mass of hadrons and nuclei verifies quantum uncertainties while that for stars and gravitational collapse verifies classical determinism.

Note also that isodeterministic principle (54) holds for the infinite class of isounitary time evolutions of the theory, Eqs. (29).

7.4. Achievement of attractive force between valence electrons.

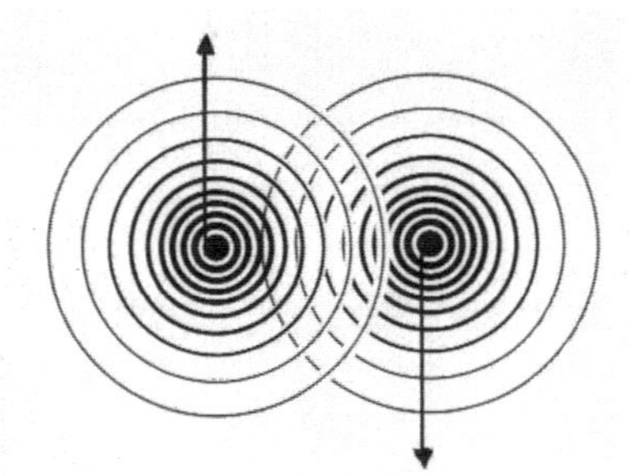

Figure 9: *A conceptual rendering of the 'attractive' force between 'identical' electrons in valence pairs first identified in Chapter 4 of the 2001 Ref. [30] against their very big Coulomb repulsion, Eq. (6), thanks to the representation of the extended character of the electron wavepackets via isomathematics and isochemistry with ensuing total EPR entanglement of the electrons. Note that this achievement proves the last statement of the EPR argument to the effect that quantum wavefunctions cannot represent 'all elements of reality' [1].*

A. Einstein, B. Podolsky and N. Rosen ended their historical paper [1] with the statement: *"While we have thus shown that the wavefunction does not provide a complete description of the physical [and we add chemical] reality, we left open the question of whether or not such a description exists. We believe, however, that such a theory is possible."*

In Section 3.5, we indicated that a primary insufficiency of quantum chemistry is the lack of an *attractive* force between the *identical* electrons in valence coupling, Eq. (5), and that their *repulsive* force is so big, Eq. (6), to prevent any realistic hope to achieve an attractive force via quantum chemistry due to its linear, local, Hamiltonian and unitary structure.

In monograph [30] of 2001, particularly Chapter 4, Santilli achieved the needed attractive force between identical valence electrons via a *non-unitary/isounitary completion of the wavefunction of quantum mechanics* which is a direct verification of the final EPR statement quoted above (see Figure 9 and Section 5 of paper [214]).

The above result can be outlined as follows. Assume that the electrons are perfectly spherical for which $n_k^2 = 1, \ k = 1, 2, 3$, and that their density is ignorable for which $n_4^1 \approx 1$. Consider then a non-unitary transformation with the following simple realization of Eqs. (45)(46)

$$\hat{I} = UU^\dagger = e^{\frac{V_{hm}}{V_{qm}}} \approx 1 + \frac{V_{hm}}{V_{qm}}$$

$$\hat{T} = (UU^\dagger)^{-1} = e^{-\frac{V_{hm}}{V_{qm}}} \approx 1 - \frac{V_{hm}}{V_{qm}} \tag{57}$$

where V_{hm} is a quantum mechanical (qm) potential, for instance, the *repulsive* Coulomb potential, and V_{hm} is a strongly *attractive* potential of hadronic mechanical (hm), such as the Hulten potential,

$$V_{qm} = +\frac{e^2}{r}$$

$$V_{hm} = -W\frac{e^{-br}}{1 - e^{-br}}, \tag{58}$$

where W is a normalization constant.

It is then easy to see that *a non-unitary transformation can turn the quantum mechanical repulsion between valence electrons into the attraction needed to represent the "element of reality" given by valence bonds*

$$UV_{qm}U^\dagger \approx V_{qm}(1 + \frac{V_{hm}}{V_{qm})} = V_{qm} + V_{hm} =$$

$$= +\frac{e^2}{r} - W\frac{e{-}br}{1 - e^{-br}} \approx K\frac{e^{-br}}{1 - e^{-br}} \tag{59}$$

where the last step is due to the fact that the Hulten potential behaves at short distances like the Coulomb potential [20]. However, the former is much stronger than the latter, by therefore allowing the absorption of the Coulomb potential into the Hulten potential irrespective of whether the Coulomb potential is attractive or repulsive. In this way, the final equation solely shows the Hulten potential with the new, positive, renormalization constant K (see page 833 of Ref. [20], Chapter 4 of Ref. [30], and paper [56]).

To achieve the corresponding iso-Schrödinger equation, recall rules (28) for the nonunitary transformation of the linear into the isolinear momentum

$$U p \psi(r) U^\dagger = U p U^\dagger (U U^\dagger)^{-1} U \psi(r) U^\dagger = \hat{p} \hat{\times} \hat{\psi}(\hat{r}) =$$

$$= -\hat{i} \hat{\times} \frac{\hat{\partial}}{\hat{\partial} \hat{r}} \hat{\psi}(\hat{r}) = -i \hat{I} \partial_{\hat{r}} \hat{\psi}(\hat{r}). \tag{60}$$

Under the above properties, the application of non- unitary transformation (57) to the conventional Schrödinger equation (5) for the valence electron pair with repulsive force then yields the iso-Schrödinger equation for their attraction as in the physical reality, which we write in its projection into the conventional Hilbert space over the field of complex numbers (see Section 2 of Ref. [214] for its detailed derivation)

$$(-\frac{1}{\bar{m}} \hat{\Delta}_r - K e^{\frac{-br}{1-e^{-br}}}) \hat{\psi}(r) = E \hat{\psi}(r), \tag{61}$$

where v$\bar{m}$ is the reduced form of isorenormalized masses, Eq. (56).

The above results confirm the final statement by Einstein, Podolsky and Rosen quoted above because the *wavefunction* $\psi(r)$ of quantum chemistry cannot represent the physical reality of the attraction between the identical valence electrons in molecular bonds in favor of its completion into the isowavefunction $\hat{\psi}(\hat{r})$ of hadronic chemistry. A more recent derivation of the attraction between valence electrons is available in Section 2.8 of paper [214].

It should be indicated that *valence electron bonds appear to be one of the most significant realizations of the EPR entanglement (Figure 1 and Section 7.2.3) apparently occurring at the limit of null mutual distance* with a considerable broadening of the implications of entanglement in physics, chemistry and biology. In fact, by merely admitting the evidence, we learn that the entanglement of wavepackets allows particle to influence each other at a distance, we learn from isotopic element (37)-(39) that entanglement plays a crucial role in nuclear forces (see also Section 8), and we now learn that, at the limit of null mutual distance, entanglement may alter the very characteristics of particles, a feature called *mutation*, which is evidently needed to turn a natural repulsion into an attraction.

It may be of some interest to note that the indicated mutation of valence electrons can be quantitatively represented by the transition from *particles* characterized by a unitary irreducible representation of the spinorial covering of the Poincaré symmetry $\mathcal{P}(.)$ into *isoparticles* characterized by isounitary isoirreducible isorepresentations of the Lorentz-Poincaré-Santilli-isosymmetry $\hat{\mathcal{P}}(.)$ [66] - [69] (see reviews Ref. [213]).

In fact, the transition from $\mathcal{P}(.)$ to $\hat{\mathcal{P}}(.)$ implies, in general, a mutation of all *intrinsic* characteristics of entangled particles also called *isorenormalization* because caused by the isotopic element representing non-Hamiltonian interactions. When the entanglement is limited, as occurring in a nuclear structure, the spin $1/2$ of the constituents is maintained, but when the entanglement is total, as occurring in valence bonds, all characteristics of the particles are expected to be mutated, including the charge.

7.5. Removal of quantum divergencies.

Recall that the divergencies of quantum mechanics originate on the singularity of Dirac's delta function

$$\delta(r) = \frac{1}{2\pi}\int_{-} \infty^{+}\infty e^{ikr}dk, \qquad (62)$$

at the origin $r = 0$.

It is easy to see that there exist an isotopic element for which the singularity at the origin $r = 0$ is removed for the *Dirac-Myung-Santilli isodelta isofunction* [45] (see also Refs. [47] - [51]), as illustrated by the following simplest possible formulation

$$\hat{\delta}(r) = \frac{1}{2\pi}\int_{-}\infty^{+}\infty \hat{e}^{ikr}\hat{d}k = \frac{1}{2\pi}\int_{-}\infty^{+}\infty e^{i\hat{T}kr}dk. \tag{63}$$

Similarly, recall that the isotopic product $A \star B = A\hat{T}B$ must be applied to the *totality* of the products, thus including all products appearing in perturbative series. But the isotopic element $\hat{T}$ has very small values in all known applications. Consequently, perturbative series that are generally divergent in quantum mechanics and chemistry are turned under isotopy into strongly convergent series. This is illustrated by fact that, given a divergent perturbative series

$$A(t) = A(0) + (AH - HA)/1! + ... \rightarrow \infty, \tag{64}$$

there always exist a value of the isotopic element $\hat{T}$ causing the strong convergence of the isotopic series

$$A(t) = A(0) + (A\hat{T}H - H\hat{T}A)/1! + ... = K \ll \infty, \ \hat{T} \ll 1. \tag{65}$$

Consequently, *the validity of Einstein's[determinism per Section 7.3 implies the removal of quantum mechanical divergencies* (see Corollary 3.7.1, page 128, Ref. [213]).

The actual verification of the above important property has been provided by D. D. Shillady and R. M. Santilli in papers [31] [32] with the proof that *the perturbative series of hadronic chemistry converge at least one thousand times faster than the corresponding quantum chemical series.*

In Santilli's view, the above rapid convergence of calculations for the EPR entanglement is an important advantage over the conventional quantum entanglement, particularly for applications such as that for computers based on the entanglement of electrons.

Note that the new computers are called in this Overview *EPR computers,* instead of 'quantum computers', due to the impossibility by quantum mechanics to provide a representation of the entanglement of electrons discussed throughout this work.

7.6. Representation of irreversible processes. As indicated in Section 3.3 and Figure 4, the insufficiency of quantum mechanics that stimulated Santilli's lifelong research on the completion of quantum mechanics according to the EPR argument [1] is the inability to represent time-irreversible events, including energy-releasing processes, due to the invariance under anti-Hermiticity of the Lie brackets for Hermitean operators , $[A, B] = -[A, B]^\dagger$.

The above occurrence stimulated the construction of the Lie-admissble branch of hadronic mechanics [6]-[9], [19]-[25], [38]-[39], with time evolution (13)-(14) (see Section 2 of Ref. [212] for a review).

Discussions during the 2020 Teleconference [0] established that genotopic elements $\hat{R}$ and $\hat{S}$ represent the external terms $F > 0$ in Lagrange's and Hamilton's equations

$$\frac{d}{dt}\frac{\partial L(t,r,v)}{\partial v} - \frac{\partial L(t,r,v)}{\partial r} = F(t,r,v),$$

$$\frac{dr}{dt} = \frac{\partial H(t,r,p)}{\partial p}, \quad \frac{dp}{dt} = -\frac{\partial H(t,r,p)}{\partial r} + F(t,r,p). \tag{66}$$

In fact, in his Ph. D. thesis of the mid 1960s [6]-[8], Santilli accepted the EPR argument [1] on the lack of completeness of quantum mechanics because quantum mechanics cannot represent the external terms F of Lagrange's and Hamilton's analytic equations which is the analytic representation of the irreversibility of nature.

Santilli noted that the brackets (A, F, H) of the classical time evolution with the external terms

$$\frac{dA}{dt} = (A, H, F) = \frac{\partial A}{\partial r}\frac{\partial H}{\partial p} - \frac{\partial H}{\partial r}\frac{\partial A}{\partial p} + \frac{\partial A}{\partial r}F, \tag{67}$$

violate the right distributivity and scalar laws, thus prohibiting the construction of a completion of quantum mechanics characterized by an *algebra* as understood in mathematics.

Therefore, Santilli turned brackets (67) into the identical but bilinear form verifying the axioms of an algebra

$$\frac{dA}{dt} = (A, H) = \frac{\partial A}{\partial r}\frac{\partial H}{\partial p} - \frac{\partial H}{\partial r}S\frac{\partial A}{\partial p},$$

$$S = 1 - \frac{F}{\partial H \partial r}. \tag{68}$$

The algebra characterized by the new brackets (A, H) turned out to be Lie-admissible according to Albert [9].

In subsequent works, Santilli studied the Lie-admissible, operator image of brackets (57) with the following realization of the genotopic elements $\hat{R}$ and $\hat{S}$

$$i\frac{dA}{dt} = (A, H) = A < H - H > A =$$

$$= A\hat{R}H - H\hat{S}A = AH - HA - AF,$$

$$R = 1, \ \ \hat{S} = 1 - F/H, \tag{69}$$

that was proved in paper [41] to be *directly universal* for the representation of all possible (non-singular) non-conservative, thus irreversible systems ("universality") directly in the coordinates of the experimentalist ("direct universality") without any need for the transformation theory.

Note that the isotopic element $\hat{T}$ is a particular case of the genotopic element $\hat{S}$ and that, in numeric values, $F < H$. Consequently, *Eqs. (69) establish the analytic origin of the important property that the isotopic element is smaller than 1,* $\hat{T} = 1 - F/H < 1$ *which property is important for the regaining Einstein's determinism.*

Additionally, *Eqs. (69) provide a quantitative identification of the insufficiency of quantum mechanics for nuclear fusions, evidently given by the absence in the quantum mechanical time evolution of the extra term AF representing irreversibility.*

It is important to clarify that, despite the lifting of Lie into Lie-admissible algebras, quantum mechanical axioms are indeed preserved. This important property can be best seen in the axiomatic representation of irreversible systems via Lie-admissible formulations whose main aspects are the following.

A tacit assumption of quantum mechanics is that the product of a quantity A times B to the right $A \rightarrow B$ is equal to the product of B times A to the left $A \leftarrow B = A \rightarrow B$ (which property is different than commutativity $AB = BA$), and the same property holds for isomechanics.

In Lie-admissible formulations, motion forward (backward) in time is represented by the axiom restricting all products to be ordered to the right $A > B$ (to the left $A < B$). The representation of irreversibility is then assured when the numeric values of the two products are different, $A > B \neq A < B$. This results in the following ordered products and units called *genoproducts and genounits to the right and to the left*

$$
\begin{gathered}
A > B = A\hat{R}B, \;\; \hat{I}^{>} = 1/\hat{R}, \\
A < B = A\hat{S}B, \;\; \hat{I}^{<} = 1/\hat{S}, \\
\hat{R} > 0, \;\; \hat{S} > 0, \;\; \hat{R} \neq \hat{S}.
\end{gathered} \tag{70}
$$

The above basic assumptions imply the duplication of isomathematics and related isomechanics, one for product to the right and one for product to the left whose review is omitted for brevity [24]. The combination of forward and backward mathematics and mechanics are known as *genomathematics and genomechanics.*

An important point is that *the abstract axioms of quantum mechanics remain fully valid for each direction of time, including the preservation of the abstract axioms of numeric fields [33], functional analysis [23] differential calculus [34], etc, merely realized in a time-ordered isotopic form.*

It is generally believed that genomathematics and genomechanics are too complex for physicists. This is unfortunate because, as shown in Ref. [35], the construction of genomathematics is truly elementary. The completion of a quantum mechanical model of nuclear fusion into a form inclusive of irreversibility, is given by subjecting all quantum mechanical quantities and operations to the following *two* non-unitary transformations U and W, including Planck's unit $\hbar = 1$, products, etc.

$$1 \rightarrow \hat{I}^{>} = U1W^{\dagger} = 1/\hat{S},$$

$$1 \rightarrow \hat{I}^{<} = W1U^{\dagger} = 1/\hat{R},$$

$$AB \rightarrow U(AB)W^{\dagger} = (UAW^{\dagger})(UW^{\dagger})^{-1}(UBW^{\dagger}) = A\hat{S}B = A > B,$$

$$AB \rightarrow W(AB)U^{\dagger} = (WAU^{\dagger})(WU^{\dagger})^{-1}(WBA^{\dagger}) = A\hat{R}B = A < B. \quad (71)$$

As illustrated in the next section, the treatment of energy releasing processes via the Lie-admissible branch of hadronic mechanics has been done via the simple, dual, non-unitary transformation (71) of quantum mechanical models.

7.7. EPR completion for matter-antimatter annihilation.

As indicated in Section 3.4 and Figure 5, the additional insufficiency of quantum mechanics studied at the 2020 teleconference [0] has been the apparent impossibility of representing the mechanism of particle-antiparticle annihilation via the conventional charge conjugation, Eq. (4).

This insufficiency lead Santilli to propose in papers [62] [63] of 1985 a fifth EPR completion of quantum mechanics, that characterizes the *isodual map* (denoted with an upper symbol d) which is also an anti-Hermitean map like charge conjugation, but it is applied to the *totality* of the quantities Q representing particles and their operations with no known exception, such as

$$Q(t, r, E, \psi, ...) \rightarrow Q^d(t^d, r^d, E^d, \psi^d, ...) =$$

$$= -Q^\dagger(-t^\dagger, -r^\dagger, -E^\dagger, -\psi^\dagger, ...). \qquad (72)$$

The above map implies that antiparticles have the *negative energy* $E^d = -E < 0$ predicted by Dirac [11] which is necessary to represent particle-antiparticle annihilation [29].

The novel mathematics characterized by isodual map (72) is known as *isodual mathematics* [29] [57], and it is based on the *isodual unit*

$$1^d = -1, \qquad (73)$$

isodual numbers [33]

$$n^d = n1^d = -n, \qquad (74)$$

isodual product

$$A \times^d B = AT^dB = -AB, \;\; 1^d = 1/T^d = -1, \qquad (75)$$

isodual spacetime

$$x^{d2d} = x^{d\mu} \times^d \eta^d_{\mu\nu} \times^d x^{d\nu} = -x^2, \qquad (76)$$

and isodual image of 20th century applied mathematics [34] [32].

On axiomatic grounds, isodual mathematics resolves the violation of causality by particles with negative energy identified by Dirac [11] because *negative energies referred to negative units are as causal as positive energies referred to positive units,* and the same holds for time and other physical quantities.

On experimental grounds, the isodual map is equivalent to charge conjugation at the particle level. Consequently, the isodual theory of antimatter represents all available experimental data in particle physics [32].

A basic novelty of the isodual theory of antimatter is that isoduality (72) predicts that antimatter emits a new light called *isodual light* [75] whose characteristics are *all* opposite to those of matter emitted light, thus implying new annihilations of the type

$$e^- + e^+ \rightarrow \gamma + \gamma^d. \tag{77}$$

which are independently requested by the symmetry of Dirac's equation and of the l. h. s. of the above decay known as *isoselfduality*, namely, the invariance under the isodual map (72) [32].

In particular, *light emitted by antimatter is predicted not to be visible by any Galileo-type refractive telescopes with convex lenses available on Earth or in terrestrial orbits,* because it requires for its detection a basically new type of telescopes currently under development. The cosmological implications of the isodual theory of antimatter are discussed in debate [76] and Ref. [163].

8. APPLICATIONS OF EPR COMPLETIONS

8.1. Foreword.

Following various requests, in this section we provide an outline of the novel applications permitted by the recent verifications [210]-[214] of the EPR argument [1] in: nuclear physics (Section 8.2), chemistry (Section 8.3), relativities (Section 8.4), and high energy scattering experiments (Section 8.5).

Since the outline presented in this section is mainly conceptual with a bare minimum of quantitative treatments, the study of the quoted literature is essential for a serious understanding of the treated topics. In addition, the literature in the field is quite vast by therefore forcing us for brevity to quote only the originating papers. Comprehensive bibliographies can be obtained from the quoted reviews.

Needless to say, the fundamental methods of this section are the *iso-, geno- and hyper-mathematics,* with particular reference to the *iso-, geno- and hyper-completions of Lie's theory,* and corresponding *iso-, geno- and hyper-mechanics and chemistry* [21]-[30]. To assist the non-initiated reader, Tutoring Lectures in the above methods have been provided in Refs. [143]-[146].

In inspecting this section, interested readers should keep in mind that the basic axioms of 20th century theories are preserved in their entirety, and merely subjected to the broadest possible realization.

Consequently, this section is intended to illustrate the remarkable broadening of the prediction and representational capabilities of the basic axioms of 20th century theories, as well as to point out that technically unsubstantiated criticisms of the applications reviewed in this section are, in reality, criticisms on 20th century theories.

Another important aspect needed for the understanding of the applications herein reviewed is that 20th century theories have one single formulation, the conventional one. By contrast, their isotopic completions have *two* formulations, that on isospaces over isofield and their *projection* on conventional spaces over conventional fields.

Recall that 20th century Hamiltonian interactions between *point-like* particles cause alterations of the characteristics of particles called *renormalization.* It is known since the original 1978 proposal to construct hadronic mechanics that contact non-Hamiltonian interactions between *extended* particles causes different alterations of the characteristics of particles called *isorenormalizations* (see Section 5 of Ref. [20], and Refs [23] [24]). The latter renormalizations are triggered by the isotopic element $\hat{T}$. Consequently, the conventional notion of *particle* is no longer applicable under EPR completion in favor of the covering notion of *isoparticles* pointed out in the preceding section [66]-[70].

As an illustration, the assumption that the particles in a valence bond are ordinary electrons leads to major inconsistencies that remain undetected by the non-initiated reader. Similarly, the assumption that the particles in the synthesis of the neutron from the hydrogen in the core of stars are conventional protons and electrons leads to inconsistencies so serious to prevent a consistent synthesis in disagreement with the experimental evidence on the existence of the neutron synthesis in nature.

8.2. Applications of EPR completions in nuclear physics.

8.2.1. Representation of nuclear magnetic moments. As it is well known, the magnetic moment of a rotating charged sphere increases (decreases) when said sphere is deformed into an oblate (prolate) spheroid while keeping constant its angular momentum. Enrico Fermi, Victor Weisskopf and other founders of nuclear physics suggested in the 1940's that the anomalous values of nuclear magnetic moments (Section 3.2 and Figure 2) may be due to a *deformation* of the charge distribution of protons and neutron caused by strong nuclear interactions.

Despite its manifestly plausible character and authoritative origination, the hypothesis of the deformability of nucleons has not yet been accepted by the mainstream physics community to date (with due exceptions) because it would require a modification of quantum mechanics essentially along the EPR argument [1]. In any case, point-like particles cannot be deformed.

The Lie-isotopic completion of quantum mechanics with two-body isounit (45), related deformation of charge distributions (Figure 10), and exact representation if the Deuteron magnetic moment first achieved in Ref. [40] of 1994, have been proposed to honor the legacy by Fermi, Weisskopf and others without altering the basic axioms of quantum mechanics.

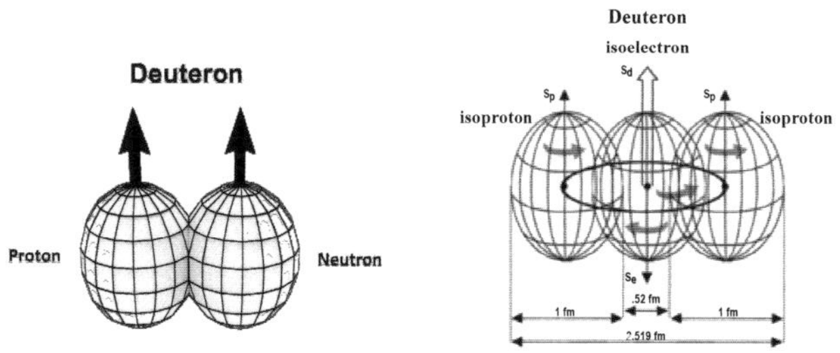

Figure 10: *The top view illustrates the deformation of the proton and the neutron in the Deuteron structure that permitted the first exact representation of the deuteron magnetic moment in Ref. [40] of 1994. The bottom view illustrates the representation of the total magnetic moment of the deuteron as a three-body hadronic bound state achieved in subsequent works [214]. Note that both views have insufficiencies for the representation of the spin* 1 *of the Deuteron that are resolved in Section 8.2.5 and Figure 13.*

However, in so doing the constituents of nuclei are no longer point-like protons and neutrons per 20th century definition, but extended, thus deformable *isoprotons* $\hat{p}^+$, *isoneutrons* $\hat{n}^0$, *or isonucleons,* as defined by the LPS isosymmetry. In fact, the 1998 proof of the EPR argument [210] applies the underlying isomathematics and isomechanics for the *exact* representation of the nuclear magnetic moment of the Deuteron $D = (\hat{p}^+, \hat{n}^0)_{hm}$, as well as of heavier nuclei, precisely along the legacy by Fermi, Weisskopf and others indicated above.

It should be noted that the deformability of nucleons was preliminarily confirmed by the 1981 neutron interferometric experiments [43] and the adjoining theoretical study on the Lie-admissible completion of the rotational symmetry [44].

Additionally, paper [210] used isomathematics and isomechanics for the reconstruction of the *exact* $\hat{SU}(2)$-isospin symmetry when believed to be broken by electromagnetic interactions. This result was achieved

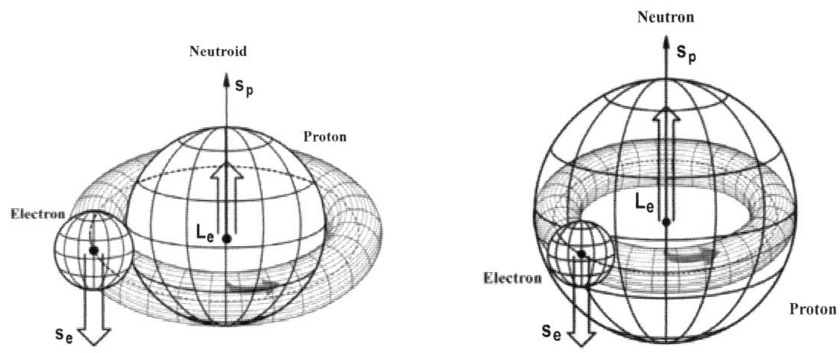

Figure 11: *At the mutual distance of* 10^{-13} *cn, the proton and the electron experience the extremely big Coulomb attraction of* 230 *Newtons, Eq. (6), although quantum mechanics prohibits their bound state. In this figure we depict the two bound states predicted by hadronic mechanics [68] [84] [214]: the neutroid in the left for a singlet coupling with total spin* 0 *and* 7 *s mean-life, and the neutron in the right with spin* 1/2 *and* 900 *s mean-life under the total compression of the electron inside the proton in the core of stars (Section 8.2.2).*

via the representation of the proton-neutron mass difference with Bohm's hidden variable λ.

The systematic reconstruction of all space-times continuous and discrete symmetries at the isotopic level when believed to be broken is presented in Refs. [23] [24].

Note that, according to the basic axioms of quantum mechanics, the triplet coupling of the proton and the neutron in the top of Figure 10 generally used to represent the spin 1 of the deuteron, is highly unstable, The only stable quantum mechanical bound state is the singlet, but in this case the total spin of the deuteron would be 0 contrary to experimental evidence. The resolution of this paradoxical situation via hadronic mechanics has been achieved in Refs. [26] [213] and will be outlined in Section 8.2.5.

8.2.2. Representation of the neutron synthesis. Nuclei are assumed to be bound states of protons and neutrons under strong interactions. However, isolated neutrons are naturally unstable. Therefore, a deeper representation of nuclear structures requires the understanding of the mechanism in which neutrons become stable when they are members of a nuclear structure. In turn, this understanding can best be achieved in the study of the neutron synthesis from a proton and an electron in the core of stars.

The problem is that the neutron synthesis is impossible for quantum mechanics despite the enormous Coulomb attraction between the proton and the electron at short distances per Eq. (6), because the mass of the neutron is $0.784\ MeV$ bigger than the sum of the masses of the proton and of the electron, in which case the Schrödinger equation would require a *positive potential energy* producing a *mass excess* which is anathema for quantum mechanics.

The 1978 Harvard University paper [20] assumed the above insufficiency of quantum mechanics as a clear evidence on the lack of completion of quantum mechanics according to the EPR argument [1] and proposed the construction of the axiom-preserving hadronic mechanics precisely for the representation of the neutron synthesis from the hydrogen.

In view of the list of insufficiencies of the conjecture that the hypothetical and directly undetectable quarks are the physical constituents of hadrons [38], paper [20] presented in Section 5 a structure model of the octet of mesons as hadronic bound states of actual physical particles produced free in their spontaneous decays with the lowest mode. Note that the model achieved compatibility with the $SU(3)$ model of the time by merely restricting it to the sole quantum mechanical *classification* of mesons, and the use of hadronic mechanics for the *structure* of individual mesons due to the known inapplicability of quantum mechanics in the interior of hadrons, much along the dichotomy classification/structure set in history for atoms and other structures.

Following the development of the Lie-Santilli isotheory [22] and its application to the isotopies of the $SU(2)$-spin symmetry [62] [63], hadronic mechanics was used in Ref. [84] for the first known non-relativistic representation of *all* characteristics of the neutron as a hadronic bound state of an isoelectron $\hat{e}^-$ and an isoproton $\hat{p}^+$

$$n = (\hat{e}^-, \hat{p}^+)_{hm}, \tag{78}$$

resulting in the prediction of *two* hadronic bound states: the first is called *neutroid* and occurs for the singlet coupling isoelectron-isoproton with spin 0 and $7\ s$ mean-life (left of Figure 11); the second is given by the neutron with spin $1/2$ and $900\ s$ mean-life, which occurs under the total compression of the isoelectron inside the isoproton (right of Figure 11).

The representation of the rest energy, mean life and charge radius of the neutron were represented by a non-unitary transformation of the conventional Schrödinger equation (1) with isotopic element of type (57) and ensuing iso-Schrödinger equation of type (61) in which the fine spectrum of the Hulten potential is constrained to have one value only, the neutron.

The representation of the spin $1/2$ required the irregular isotopies $\hat{SO}(3)$ and $\hat{SU}(2)$ to represent the *hadronic angular momentum* of the isoelectron when compressed inside the hyperdense isoproton, thus being forced to rotate with its spin, resulting in an *orbital* eigenvalue $1/2$ which is impossible for the *unitary* $SU(2)$ symmetry of quantum mechanics, but readily possible for the *non-unitary* $\hat{SU}(2)$ isosymmetry.

The representation of the anomalous magnetic moment of the neutron became possible thanks to the contribution from the rotation of the isoelectron inside the isoproton which is completely absent in quantum mechanics (see the review and upgrade in paper [214]).

The relativistic representation of *all* characteristics of the neutron in its synthesis from the hydrogen was achieved in the 1993 paper [68] written at the Joint Institute for Nuclear Research, Dubna, Russia (see also the 1995 paper [69] published in China).

By ignoring in first approximation the negative binding energy due to the Coulomb interactions between the isoelectron and the isoproton (because the Coulomb attraction is absorbed by the Hulten potential), iso-Schrödinger equation (62) admits physically meaningful solutions for the Hulten potential only when the isorenormalized rest energy of the isoelectron, Eq. (36), acquires the value

$$\bar{m}_{\hat{e}} = \frac{m_e}{n_4^2} = 0.511/n_4^2 = 1.295\ MeV, \tag{79}$$

from which we obtain the density value

$$n_4^2 = 0.394, \tag{80}$$

which is fully within the values of similar densities, such as that of the proton-antiproton fireball obtained from the fit of the experimental data of the Bose-Einstein correlation, which are given by the value $n_4^2 = 0.428$ of Ref. [85] , Eq.(10.27a), page 127 and the value $n_4^2 = 0.364$ of Ref. [86], Table I, page 441.

As recalled earlier, the radial equation of iso-Klein-Gordon equation (35) or the iso-Dirac equation (163), do not admit physically meaningful solution in the event the rest energy of the neutron is *smaller* than the sum of the rest energies of the electron and of then proton, thus suggesting isorenormalization (79) that includes the missing $0.782\ MeV$.

A primary function of the iso-Minkowskian geometry [70] is the characterization of the isorenormalization of the rest energy of the electron caused by non-Hamiltonian interactions in the interior of the neutron in such a way to avoid the indicated "mass-energy defect" and allow isorelativistic equations to have physical solutions.

Needless to say, the above relativistic effects mandate the study of the laws applicable in the hyperdense medium inside the neutron, which study is rudimentarily done in Section 8.4.4, see in particular Isoaxiom IV, Eqs. 133 (see alsoVolume IV or Ref. [87] for a 2008 review and Ref. [214] for a recent update).

Note that the interactions with representation (45) which bond together the proton and the electron into the neutron constitute an explicit form of *fifth interactions* according to its definition in Section 7.2.3.

In Section 8.2.5, we shall outline the significance of the neutron synthesis for the representation of nuclear data, such as the stability of nuclei as bound state of protons and neutrons despite the natural instability of the neutron when isolated.

8.2.3. Etherino or neutrino?] Due to the respect toward the memory of W. Pauli and E. Fermi, it took years to indicate that in both, the non-relativistic [84] and relativistic [68] derivations, there is no energy available for the neutrino due to the fact that the neutrino is written in the *right-hand-side* of the synthesis,

$$p^+ + e^- \ \rightarrow \ n + \nu, \tag{81}$$

while $0.784\ MeV$ are missing for the synthesis of the neutron itself.

Additionally, the isotopic $\hat{SU}(2)$-spin isosymmetry established that there is no spin available for the neutrino because the total spin of the isoelectron inside the neutron must be 0. This is due to the fact that the orbital motion of the isoelectron inside the isoproton must be equal to the isoproton spin to avoid huge resisting forces and be opposite to the isoelectron sped for a stable singlet coupling (right of Figure 11).

These stability conditions confirm the original conception of the neutron synthesis by Rutherford in 1920 [88],

$$\hat{p}^+ \hat{e}^- \ \rightarrow \ n. \tag{82}$$

In summary, despite a widespread oblivion in the best Ph. D. schools of physics, the neutron synthesis is indeed the most fundamental event in nature with yet unsolved basic problems. For instance, it is generally believed that the missing energy of $0.784\ MeV$ is provided by the latent energy in the interior of stars. However, at the initiation of the production of light, our Sun synthesizes 10^{38} neutrons *per second*. In the event the missing energy is provided by the latent energy, stars would *lose* $10^{38}\ MeV$ per seconds, thus cooling down and not being able to produce light.

It is also believed that the missing energy in the neutron synthesis is provided by the $p^+ - e^-$ relative kinetic energy. However, the scattering cross section of protons and electrons at about $1MeV$ is so small, $10^{-20}\ Barns$, to prohibit any synthesis.

Independently from the above aspects, in supernova explosions we have the emission of many times the entire energy released by the Sun in its entire life of 10 billion years. Calculations have shown that the emission of such enormous amount of energy in one single explosion cannot be consistently represented via the sole admission of quantum mechanical processes.

In view of the above, paper [89] of 2007 suggested that the missing energy in the neutron synthesis is provided by the *ether* as a universal substratum with extremely high energy density (to propagate light at $300K\ km/s$) via a longitudinal, spin 0 impulse called *etherino*, (denoted with the letter "a" from the Latin *aether*) and placed in the *left-hand-side* of the synthesis,

$$p^+ + a + e^- \rightarrow n. \tag{83}$$

It appears that the experimental data of *indirect events* predicted by the neutrino hypothesis can be numerically interpreted via the etherino hypothesis, of course, jointly with corresponding reinterpretations in the standard model for its sole use for the classification of particles.

The historical criticism against the ether that it would violate special relativity has been long dismissed because we would never be able to identify a reference frame at rest with the ether. The additional historical criticism against the ether as a physical medium given by the *aethereal wind* that should be experienced by moving masses, has also been debunked long time ago (see the 1956 paper [90]) because, from the known law $E = h\nu$, the electron is characterized by oscillations with the frequency of

$$Hz_e = 1.25 \times 10^{20} Hz. \quad (84)$$

The idea that the above oscillations are those of a tiny mass inside the electron has no scientific credibility. Hence, the sole known plausible hypothesis is that *the electron is characterized by oscillations of one point of the ether with frequency* (84). When the electron moves, said oscillations are transferred to different points of the ether with no possible "aethereal wind," and the same holds for all particles and, therefore, for matter. Inertia is the resistance by the ether against changes in the transfer speed of the oscillations to new points.

Note that, far from being pure philosophical considerations, the problem of the structure of the electron is crucial to achieve an understanding of the mechanism according to which two identical valence electrons bond themselves in molecular structures against their huge Coulomb repulsion (see Section 8.3.1).

Note also that the *transverse* character of electromagnetic waves (oscillations perpendicular to the direction of motion) mandates that the ether should have a "rigid-type" medium, while, by contrast, a "fluid-type" structure of the ether would be irreconcilable with the transverse characteristics of light.

To illustrate the implications of the problem here considered, the above view on the ether would imply that *matter is completely empty and space is completely full* to such an extent that, in case we could stop time, the entire universe would disappear.

Also, the etherino hypothesis provides a concrete example of the historical hypothesis of the *continuous creation of matter in the universe,* occurring not only for the neutron synthesis in the core of stars, but also for supernova explosions and other possible events.

Recall that, according to available measurements, the Sun loses via radiation the rather sizable amount of

$$\Delta M_{sun}^{loss} = -4.26 \times 10^9 \ Kg \ per \ second. \tag{85}$$

In the event the etherino hypothesis is correct, the above loss would be mostly counterbalanced by the synthesis in the Sun of 10^{38} neutrons per second implying an *increase* of its mass by about [163]

$$\Delta M_{sun}^{gain} = +2.87 \times 10^8 \ kg \ per \ second, \tag{86}$$

resulting in the rather small loss by the Sun of about $3f73 \ Kg/s$, which is well within current errors, plus secondary contributions, such as accretion of particles during travel in intergalactic space.

Rather than rejecting the above view due to its novelty, an important scientific question is whether current knowledge in planetary trajectory do set an upper value on the possible loss of mass by the Sun per second (additional aspects are treated in Ref. [163] and in the EPR Debate [91] and Ref. [136]).

8.2.4. Industrial synthesis of neutrons from the hydrogen. Tests on the laboratory synthesis of the neutron from the hydrogen were initiated in the late 1940's with support by Albert Einstein, but the related papers were never accepted for publication by scientific journals of the time. During the last of these initial tests, don Carlo Borghi detected nuclear transmutations apparently caused by particles he called *neutroids* emitted by the apparatus without the direct detection of synthesized neutrons (see the historical account [92]).

Following, and only following, the theoretical understanding of the neutron synthesis via hadronic mechanics (Section 8.2.2.), systematic tests on the laboratory synthesis of the neutron were initiated in the early 2000's. The first actual detection of the emission of neutrons synthesized

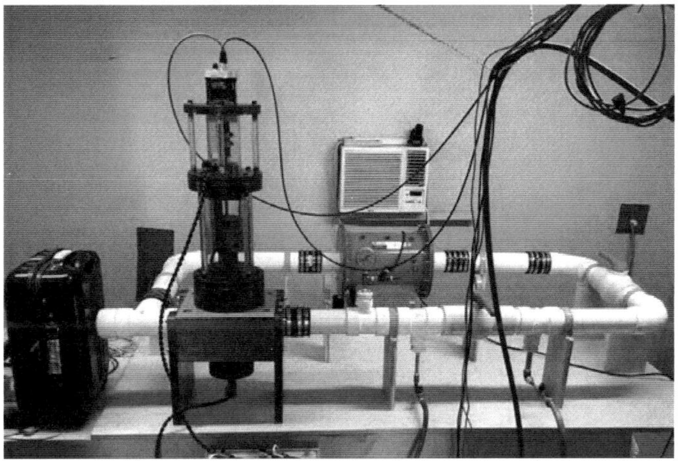

Figure 12: *A view of the Directional Neutron Source (DNS) described in Section 8.2.4 which synthesizes on demand neutrons from the hydrogen in the desired direction, flux and energy.*

from a hydrogen gas were presented in paper [93] of 2007, with the repeated detection of synthesized neutrons as well as the confirmation of the synthesis of don Borghi's neutroids via their stimulated nuclear transmutations (see paper [214] for a recent review).

The neutron synthesis was achieved via the use of a special electric arc submerged within a hydrogen gas which: ionizes the gas, synthesizes the neutroid and then synthesizes the neutron by "compressing" the isoelectron inside the isoproton much according Rutherford's original conception [88].

These tests led to the formation of the U. S. publicly traded company *Thunder Energies Corporation* (now Hadronic Technologies Corporation) for the manufacturing and sale of equipment called *Directional Neutron Source* (DNS) which produces on demand a neutron beam synthesized from the hydrogen gas in the desired direction, flux and energy (see Figure 12, the latest systematic tests [94] and the lectures by S. Beghella Bartoli in the 2020 Teleconference [0]).

Subject to appropriate funding, the DNS is intended to: 1) Provide the clear detection of nuclear weapons or materials smuggled in luggage; 2) Detect precious metals in mining operation; 3) Test weldings of thick plates in naval constructions; and other applications.

8.2.5. Representation of Deuteron data. As indicated in Section 3.2, a second important verification of the EPR argument in nuclear physics is the inability by quantum mechanics to represent the Deuteron magnetic moment, spin, stability and other data *for isolated Deuterons in their ground state.*

Recall from Section 8.2.2 the structure model of the neutron as a hadronic bound state of an isoproton with spin $S_p = 1/2$ and an isoelectron compressed in its interior in singlet coupling with total spin $\hat{J}_e = \hat{S}_e + \hat{L}_e = 0$ in which the hadronic angular momentum is constrained to coincide with the spin of the isoproton $\hat{L}_e = \hat{S}_p = 1/2$ to prevent extreme resistive forces due to the *extended* character of all particles.

Preliminary studies ion the Deuteron structure as a *three-body* bound states of two isoprotons and one isoelectron have been presented in Section 2.5, page 181 on, of Ref. [26] and in subsequent works (see Refs. [212] and [214]). However, the resulting structure models turned out to have insufficiencies for the consistent representation of the Deuteron spin $J_D = 1$ due to the lack of interior orbital conditions for the isoelectron to assume the null value $\hat{J}_e = \hat{S}_e + \hat{L}_e = 0$ (see Figure 10 for details).

We here present, for the first time, the apparent resolution of the above insufficiencies via the *axial triplet coupling,* of one isoproton and one isoneutron which occurs for spin parallel along the same symmetry axis, as illustrated in the Figure 13

$$D = (\hat{p}^+_\uparrow, \hat{n}^0_\uparrow)_{hm} = (\hat{p}^+_\uparrow, \hat{e}^-_0, \hat{p}^+_\uparrow)_{hm}, \tag{87}$$

where the first expression represents the original configuration, while

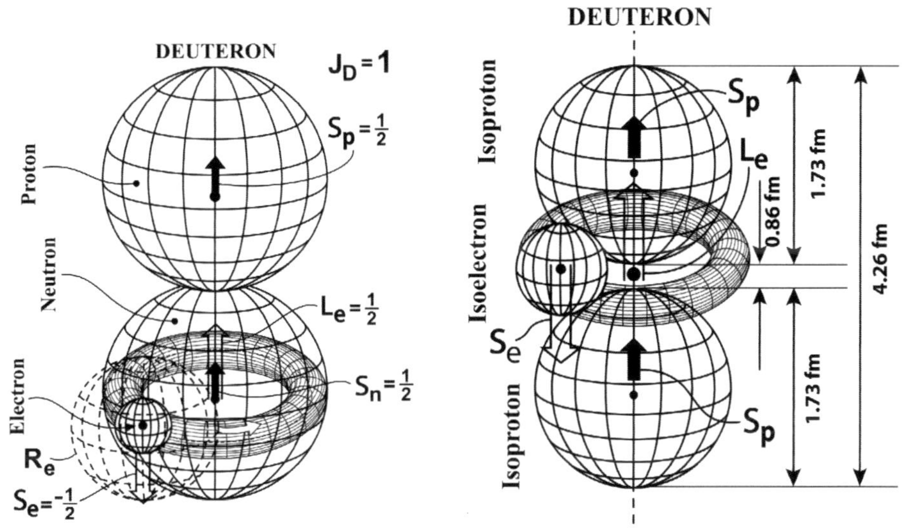

Figure 13: *Quantum mechanics predicts that the stable bound state of a proton and a neutron is the singlet, in which particles are coupled with opposing spins. The representation of the Deuteron spin* $J_D = 1$ *then requires the admission of a number of orbital states, against the experimental evidence that the value* $J_D = 1$ *occurs n the ground state. The consistent representation of the spin* $J_D = 1$ *of the Deuteron in the ground state has been achieved, apparently for the first time, in this Overview as follows. Under the assumption that the neutron is a hadronic bound state of a proton and an isoelectron (Section 8.2.2), the most stable bond state of a proton and a neutron is the "axial triplet coupling" illustrated in this figure where the left view illustrates the initial configuration, and the right view illustrates the final configuration in which the negatively charged isoelectron acts like a "gluon" between the two positively charged protons, by permitting a consistent representation of all Deuteron data (see Section 8.2.5 for details).*

the second expression represents the final configuration of the negatively charged isoelectron acting like a "gluon" between the two positively charged isoprotons. The "hat" in all particle symbols denotes deep EPR entanglement of the particles under contact non-Hamiltonian interactions.

The importance of the axial triplet coupling of Figure 13 is that it allows the intermediate isoelectron to have an *interior orbital motion within the two isoprotons* with $L_e = S_p = 1/2$ and $J_e = 0$, thus representing the total spin $J_D = 1$ *in the ground state.*

It may be of some interest to note that the representation of the Deuteron spin $J_D = 1$ has been controversial for about one century, because quantum mechanics predicts that the stable bound state of a proton and a neutron is the singlet for which $J_D = 0$ (see the top view of Figure 10)

$$D = (p_{\uparrow}^{+}, n_{\downarrow}^{0})_{qm}. \tag{88}$$

Since the total spin of such a bound state is $J_D = 0$, the salvage of quantum mechanics has been attempted via the controversial assumption of a collection of *orbital states* yielding the needed total value $J_D = 1$ against the experimental evidence that isolated Deuterons are in their *ground state.*

The Deuteron structure model as a hadronic bound state of one isoproton and one isoneutron in axial triplet coupling appears to resolve this century old controversy.

A reason for the preference of the axial triplet coupling neutron-proton with $J_D = 1$, compared to the singlet coupling with $J_D = 0$, seems to be due to the large spheroid comprising the wavepacket of the isoelectron with radius of about 2×10^{-10} cm which is solely present in the structure of theneutron (see the dashed circle denoted R_e in the left of Figure 13). This implies a geometric inequality between the isoproton and the isoneutron which clearly favors the axial triplet coupling over the singlet coupling.

Under the above assumptions, model (87) allows the representation of *all* experimental data of the Deuteron *in its true ground state* as follows (see Section 7 of paper [214] for details):

1) The representation of the total spin $J_D = 1$ of the Deuteron by apparently confirming that *the Deuteron is a three-body bound state.*

2) The representation of the anomalous magnetic moment of the Deuteron thanks to the spin and orbital contributions of the isoelectron to the total magnetic moments.

3) The representation of the stability of the Deuteron despite the natural instability of the neutron (when isolated) because isoprotons and isoelectrons are permanently stable particles.

4) The representation via the iso-Schrödinger equation of the rest energy and size of the Deuteron under the assumption that *strong nuclear interactions are represented by the isotopic element $\hat{T}$ of Eqs. (38)(39).*

5) The representation of nuclear exchange forces via the exchange of the central isoelectron by the two adjacent isoprotons.

Note that the above results are primarily due to the *explicit and concrete representation of strong nuclear interactions via the isotopic element $\hat{T}$ with realization (45).*

Note also that strong nuclear interactions with representation (45) additionally provide a concrete and explicit realization of Bohm's *hidden variables,* and characterize *fifth interactions* according to their definition given in Section 7.2.3.

A. A. Bhalekar and R. M. Santilli used the above results for the Deuteron to achieve a representation of the magnetic moment and spin of stable nuclei at large [95].

8.2.6. Reduction of matter to isoprotons and isoelectrons. The preceding results, with particular reference to the reduction of the neutron to a hadronic bound state of an isoproton and an isoelectron, as well as the ensuing representation of nuclear data in their ground state, imply the reduction of all matter in the universe to isoprotons and

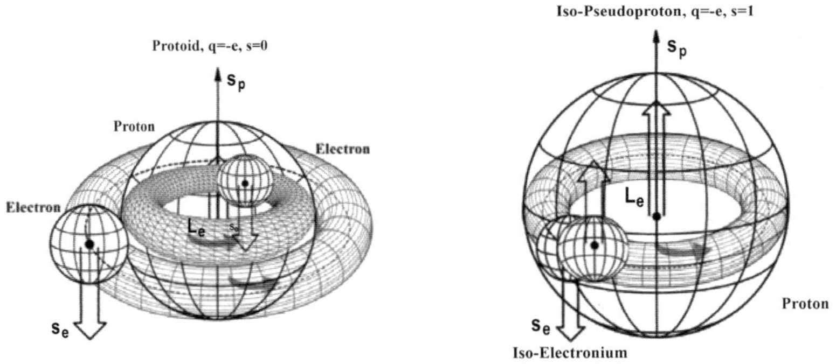

Figure 14: *We here provide an illustration of the following two negatively charged strongly interacting particles predicted by the EPR completion of quantum into hadronic mechanics with mean-lives of the order of seconds and mass close to that of the neutron: the protoid (on the left) with spin* 0, *which is a hadronic singlet state of an electron and a neutron, and the pseudo-proton (on the right) which occurs when the electron is totally compressed inside the neutron with spin* 1 *(Section 8.7). Note that the indicated new particles have been produced by the Directional Neutron Source of Section 8.2.4 operating on a commercially available Hydrogen gas (See Section 8.2.7 for details).*

isoelectrons, including protons and electrons as a particular case [26]. Note that neutron stars are equally reduced to isoprotons and isoelectrons.

8.2.7. The synthesis of the pseudo-proton. Following the synthesis of the neutron via the compression of an electron within the hyperdense proton, the Directional Neutron Source (DNS) of Section 8.2.4, progressively synthesizes, in a statistical smaller amount, *two negatively charged strongly interacting particles.*

The first particle is predicted to be characterized by a hadronic singlet coupling of an isoelectron and an isoneutron called *protoid* and denoted

$\bar{p}_1^-$ [96] with spin 0, mass essentially that of the neutron and mean-life predicted to be of about $5\ s$ (left of Figure 14). The existence of the protoid is predicted from the fact that non-Hamiltonian interactions caused by deep EPR entanglement according to structure equation (61) are insensitive to charges .

The second particle, called *pseudo-proton* and denoted $\bar{p}_2^-$ [96] is predicted from the compression of the electron, this time, inside the hyperdense neutron, (right of Figure 14) with spin 1, mass essentially that of the neutron and mean-life of the order of $7\ s$ (see also [214] for a recent review).

In regard to the spin of the pseudo-proton, we recall that electrons couple in singlet bonds. Therefore, when an electron is compressed inside the neutron, it couples in singlet with the pre-existing internal electron, resulting in a electron pair with total spin $J = 0$ which is *constrained* to rotate by the hyperdense medium inside the neutron with the neutron spin $J = 1/2$, resulting in the total spin $J = 1$.

Different models are faced with the extreme resistance that would be experienced by an extended wavepacket to rotate inside and against the rotation of the hyperdense neutron medium.

The emerging new technology, called *pseudo-proton irradiation* is made possible by the fact that, even though the pseudo-proton is evidently unstable, its mean-life is nevertheless of the order of a few seconds (a fraction of the $900\ s$ mean life of the neutron), thus being suitable for applications.

By noting on industrial ground that *pseudo-protons are strongly attracted by nuclei,*pseudo-proton irradiation is significant for the development of new clean nuclear energies, for the recycling of nuclear waste, and for other intriguing applications studied in Section 8.2.10.

On medical grounds, *pseudo-proton irradiation*i is under consideration for cancer treatment via a localized low energy beam in replacement of proton irradiation due to its excessive invasive character caused by the high energies needed to overcome proton-nucleus repulsion.

It should be noted that the existence of the pseudo-proton, let alone its long mean life for particle standards, is impossible without the EPR completion of quantum into hadronic mechanics.

8.2.8. The synthesis of the pseudo-Deuteron. The Directional Neutron Source (DNS) of Section 8.2.4, when filled up with a commercial grade hydrogen gas, synthesizes neutrons and pseudo-protons (as well as their intermediate states).

When filled up with Deuterium gas, the same DNS is predicted to synthesize a *negatively charged nucleus* called the *pseudo-Deuteron* [96] [214] according to the structure

$$\hat{D} = [\hat{p}^+_\uparrow, (\hat{e}^-_\downarrow, \hat{e}^-_\uparrow), \hat{n}^0_\uparrow]_{hm}, \tag{89}$$

namely, via the compression of an electron pair $(\hat{e}^-_\downarrow, \hat{e}^-_\uparrow)$ inside the Deuteron $D = (p^+_\uparrow, n^0_\uparrow)$, resulting in a new *negatively charged* nucleus that (in nuclear symbols A = atomic number, Z = charge and J= spin) is denoted $\hat{D}(-1, 2, 1)$ with evident decay $\hat{D} \to D + 2\beta$.

To understand the above synthesis, let us recall that the valence electrons of the Deuterium molecule are generally bonded into the valence pair $(\hat{e}^-_\downarrow, \hat{e}^-_\uparrow)$ called *isoelectronium* [30] (Section 8.3.1) with charge $2\,e$, total spin 0, total angular momentum 0 and total magnetic moment 0.

The specially designed DC arc of the DNS separates Deuterium molecules by forming a plasma mostly composed by protons, electrons and valence pairs (Section 8.1). Next we should recall the *very big Coulomb attraction between valence electron pairs and natural Deuterons* according to Eq. (6). Such an attraction evidently prepares synthesis (89) by forming an intermediate state called *Deuteroid* [96]. The same, specially designed DC arc then compresses the valence pair inside the Deuteron according to the configuration of Eq. (89) and Figure 15.

This process creates a negatively charged nucleus which is evidently

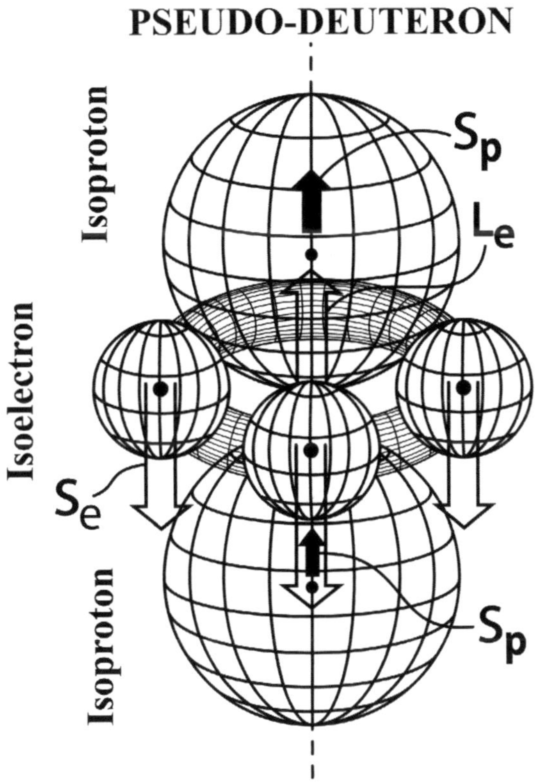

Figure 15: *An illustration of the negatively charged nucleus, called pseudo-Deuteron and denoted $\hat{D}(-1,2,1)$, according to structure (89), which is given, firstly, by the very big Coulomb attraction between electron pairs and Deuteron nuclei and, secondly, by the compression of a valence electron pair inside a natural Deuteron permitted by the sufficiently powerful DC electric arc of the DNS.*

unstable, but its main-life (computed via hadronic mechanics) is of the order of a *second*, thus allowing industrial applications (see Section 8.2.9-III). Needless to say, the plasma created by the DC arc submerged within a Deuterium gas, also contains isolated electrons that are equally attracted by Deuterons, and can form other hadronic bound states with smaller statistical probabilities and mean-lives [96].

8.2.9. New clean nuclear energies. In this section, we provide a conceptual outline of new nuclear energies without harmful radiations called *hadronic energies* originally proposed in paper [97] of 1994 , which are not possible for quantum mechanics, but they are indeed possible for the completion of quantum into hadronic mechanics [23]-[25] according to the Einstein-Podolsky-Rosen argument [1] following the exact representation of nuclear spin, magnetic moments, stability and other data presented in the preceding sections.

The *existence* of the new energies in parts per million (ppm) has been experimentally verified by various independent laboratories thanks to funds provided by Magnegas Corporation (now Taronis Corporation), but no funds could be located for their development up to the level of *industrial* production of new nuclear energies, which development is therefore left to interested scientists and governmental officers.

Among the variety of hadronic energies predicted in Ref. [97] (thanks to the Lie-admissible completion of quantum mechanics for irreversible processes) we outline the following three clean energies (see Refs. [25] and [98] for reviews):

8.2.9-I. Nuclear energies via stimulated neutron decays. Incontrovertible experimental evidence establishes that the neutron is synthesized from the proton and the electron and decays into the proton and the electron. Hadronic mechanics has permitted the representation of *all* characteristics of the neutron at both non-relativistic and relativistic levels (Section 8.2.3) when assumed to be a hadronic bound state of an isoproton $\hat{p}^+$ with conventional rest energy of $938.272\ MeV$ and an isoelectron $\hat{e}^-$ with isorenormalized rest energy of $1.295\ MeV$, Eq.(80) due to its immersion within the hyperdense medium inside the proton with density (79).

In view of the above data, Ref. [97] submitted the hypothesis that *when the member of a suitably selected, light, natural, and stable nucleus, the neutron can be stimulated to decay* via a resonating frequency which is an integer multiple or submultiple of $\hat{\gamma} = 1.295\ MeV$ corresponding to the resonating frequency (see Eq. (3.6), page 326, Ref. [87])

$$\nu_{reson} = 3.1289 \times n \times 10^{20}\ Hz,\ n = 1, 2, 3, ... \tag{90}$$

or

$$\nu_{reson} = 3.129/n \times 10^{20}\ Hz. \tag{91}$$

By recalling that $1.295\ MeV = 0.0014\ amu$, and by ignoring integer multiples or submultiples for a first analysis, the indicated hypothesis can be written

$$\hat{\gamma}(0, 0, 1, 0.0014\ amu) + N(Z, A, J, amu) \rightarrow$$

$$\rightarrow \hat{N}(Z + 1, A, J + 1, amu) + \hat{e}^-(-1, 0, 0, 0.0014\ amu). \tag{92}$$

Note that the energy supplied by the resonating photon is reemitted by the isoelectron which has the rapid spontaneous decay in vacuum

$$\hat{e}^- \rightarrow e^- + \nu\ (or\ a), \tag{93}$$

where the etherino alternative a is outlined in Section 8.2.4.

Among various possible hadronic energies, Ref.[97] suggested in page 340 the test of the stimulated decay of a neutron of $30 - Zn - 70$ according to the reaction (see Ref.[99] for nuclear data)

$$\hat{\gamma} + Zn(30, 70, 0, 69.9253) \rightarrow$$

$$\rightarrow \hat{Ga}(31, 70, 1, 69.9253) + \hat{\beta}^-, \tag{94}$$

where $\hat{Ga}$ is a highly unstable pseudo- nuclide due to the missing energy of $0.0007\ amu$ to have the tabulated mass for $31 - Ga - 70$ of $69.9260\ amu$, with spontaneous decay

$$\hat{Ga}(31, 70, 1, 69.9253) \rightarrow Ge(32, 70, 0, 69.9242) + \beta^- + \nu, \tag{95}$$

where $32 - Ge - 70$ is a stable light natural element. Note that the indicated transmutation triggered by the stimulated neutron decay provides *two* different new clean hadronic energies without the emission of harmful radiation or release of radioactive waste: 1) Heat produced in the amount of $0.0011\ amu = 1.024\ MeV$ per reaction, and 2) Electricity generated by the production of two electrons per reaction which can be easily trapped via a thin metal shield to form a *nuclear battery* (see Sections 4-I and 4-II of Ref. [97] for details.).

The stimulated double beta decay was tested in Ref. [100] (see also the subsequent presentation [121]) with preliminary positive results reported in Ref. [87].

The tests can nowadays be done by irradiating a plate of the selected isotope (e.g., $30 - Zn - 70$) with resonating frequency available from radioactive isotopes for a few days, and then have comparative mass spectroscopic analyses of the untreated and the treated plate to ascertain the possible presence in the treated plate of the predicted new isotope $32 - Ge - 70$ in ppm.

The direct costs for the repetition of the tests are approximately given by: 1) $1,200 for a sample of the needed pure $30 - Zn - 70$ isotope with certified content; 2) $2,200 for the purchase of the needed radioactive isotope; and 3) $800 for mass spectroscopic analyses, for the total direct cost of $4,200$.

It should be noted that there exist a considerable amount of scientific literature on double beta decays (which is easily identifiable via an internet search), although none of them, to our knowledge, considers the *stimulated* double beta decay proposed in Ref. [97] of 1994 and tested in Ref. [100] of 1996. The study of the triggering of known double beta decays with resonating photons to search for new clean energies is left to interested colleagues.

8.2.9-II. Intermediate Controlled Nuclear Fusions. It is generally believed in mainstream physics circles that *nuclear fusions cannot occur without the emission of neutrons.* It is important for our environment to disregard such a view because generally based on the dismissal of the insufficiencies of quantum mechanics in nuclear physics (Section 3.2), and expressed in oblivion of Einstein's view on the limitation of quantum mechanics. This section is devoted to physicists interested in testing nuclear fusions without the emission of neutrons, as predicted by the completion of quantum into hadronic mechanics according to the EPE argument.

Additionally, this section is intended for physicists interested in verifying or dismissing in refereed scientific publications that the sole scientific, i.e., quantitative interpretation of thunder is that via nuclear fusions of light, natural and stable elements triggered by lighting without the emission of neutrons.

As it is well known, the *existence* of nuclear fusions (also called syntheses) at low energies has been established by the so-called *cold fusions,* but the energy available turned out to be insufficient to power all engineering means needed for industrial production of clean energy. The existence of nuclear fusions has also been established by the so-called *hot fusions,* but their energies and temperatures are so big to create uncontrollable instabilities at the time of the initiation if the fusion process.

Despite the investment of billions of dollars of public funds over half a century, cold and hot fusions have been unable to achieve the much needed, industrially viable, clean nuclear energies.

The apparent primary reason is that all research in the field was based on quantum mechanics in oblivion of the EPR argument [1] as well as of the known impossibility for quantum mechanics to provide an exact representation of nuclear data, as well as the impossibility to provide a correct representation of energy releasing processes due to their time irreversibility (Section 3.3).

In view of the above occurrence, systematic research was initiated via funds provided by Magnegas Corporation in the search of new clean nuclear energies based on the completion of quantum mechanics into the irreversible, Lie-admissible branch of hadronic mechanics. This research can be briefly outlined as follows.

Following, and only following the achievement of sufficient mathematical [23] and physical [24] maturity on he Lie-admissible branch of hadronic mechanics (see also lecture [101]), a study was conducted on the formulation of the *new* physical laws of nuclear fusions according to hadronic mechanics, which laws were presented in the 2008 paper [102] to characterize the new *Intermediate Controlled Nuclear Fusions* (ICNF), namely, nuclear fusions with energy intermediate between the cold and hot fusions.

The main differences between quantum fusions and ICNF are: 1) Nuclei are massive points for quantum fusions, while nuclei are extended for ICNF; 2) Quantum fusions solely admit action-at-a-distance potential interactions, while ICNF are primarily based on contact non-potential interactions; and 3) Quantum fusions are solely based on physical processes, while ICNF are based on physical as well as certain crucial chemical processes indicated in Section 8.3, thus requiring both hadronic mechanics and chemistry (see the review in Section 3.25 of Ref. [25]).

A hadronic reactor for the test of ICNF was built in 2009 comprising:

1) A metal vessel housing in its interior a pair of Carbon electrodes whose gap is electronically controlled from the outside;

2) A selected light, natural and stable gas called *hadronic fuel* which is stored at pressure in the metal vessel;

Figure 16: *A view of the hadronic reactor used for the Intermediate Controlled Nuclear Fusion (ICNF) of the Nitrogen from Deuterium and Carbon without the emission of neutronic or other harmful radiations first achieved in Ref. [103] of 2010.*

3) A DC arc between a pair of electrodes submerged within the hadronic fuel powered by a commercially available $50\ KW$ DC welder produced by Miller Electric corporation which delivers up to $1K\ A$ in the arc between the submerged electrodes with about $7\ mm$ gap;

4) Means for the control of the internal nuclear fusions including the control of the DC power, gas pressure, electrode gap, and other engineering means;

5) The hadronic reactor is then completed with a number of peripheral equipment, such as: a variety of radiation detectors with alarms preset at minimal reading for safety; vacuum and pressure pumps; vacuum and pressure gages; internal and external temperature gages; electric panel for the monitoring and control of the pressure, temperature, radiation; and other equipment (see Figure 18).

Samples of the gaseous hadronic fuel are taken before and after the activation of the reactor whose operation is limited to a few minutes for evident safety due to a generally rapid increase of the temperature. ICNF

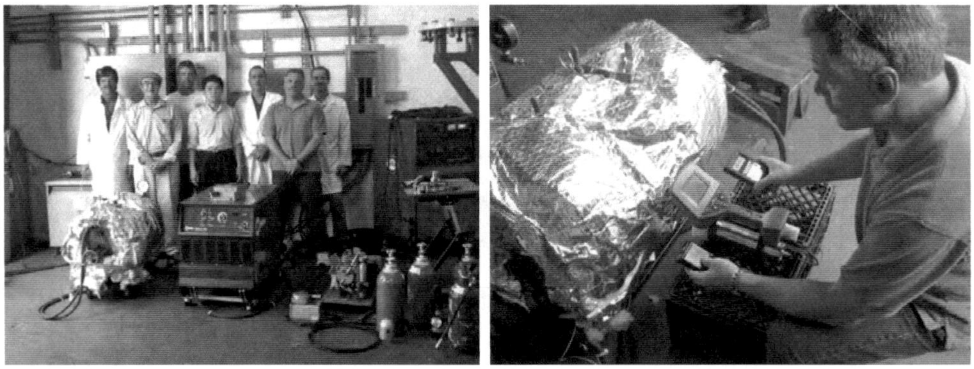

Figure 17: *A view of the team of scientists who confirmed the ICNF of the Deuterium and Carbon into the Nitrogen without the emission of harmful radiations and without the release of radioactive waste [100]-[112].*

occur when laboratory analyses signed by the laboratory director establish the presence of new light, natural and stable elements at least in ppm, under the condition that said new elements do not exist in the original hadronic fuel.

Following extensive tests, Ref. [103] of 2010 announced the ICNF in ppm of the Nitrogen from the Carbon of the electrodes and a commercial grade Deuterium gas as hadronic fuel (Figure 16) according to the ICNF

$$D(1,2,1^+,2.0141) + C(6,12,0^+,12.0000) \rightarrow$$
$$\rightarrow N(7,14,1+,14.0030) + \Delta E_{heat}, \quad (96)$$

confirmed by chemical analyses [104]-[107] signed by the laboratory director, where the released heat is given by

$$\Delta E_{heat} = 10.2231\ MeV = 1.57 \times 10^{-15} BTU \quad (97)$$

per fusion.

Figure 18: *A view of the third hadronic reactor for the ICNF of the Carbon and Oxygen into the silicon successfully tested in 2011 [109].*

It should be stressed that the indicated Nitrogen synthesis can only occur **without any production of neutron or other harmful radiation or release of radioactive waste** since it deals with the synthesis of two light, natural and stable elements into a third light, natural and stable element with a smaller mass, as well as due to the limited power of the DC source (50 KW) which is clearly insufficient to crack nuclei for the production of the popularly expected neutrons. A recording of the sound of the hadronic reactor during operation is available from Ref. [108].

The above results were all confirmed by systematic tests and verifications [109] - [112] conducted by an independent team of scientists (Figure 17).

Among a number of additional hadronic reactors build for the test of ICNF, we mention the third hadronic reactor (Figure 15) built in 2011 for a Chinese client for the ICNF of Carbon and Oxygen into the Silicon,

$$C(6,12,0^+,12.0000) + O(8,16,0^+,17.9991) \rightarrow$$
$$\rightarrow Si(14,28,0^+,27.9769) + \Delta E, \qquad (98)$$

were

$$\delta E = 2.0222\ amu = 1,882.6682\ MeV = 2.8604 \times 10^{-13} BTU/reaction, \qquad (99)$$

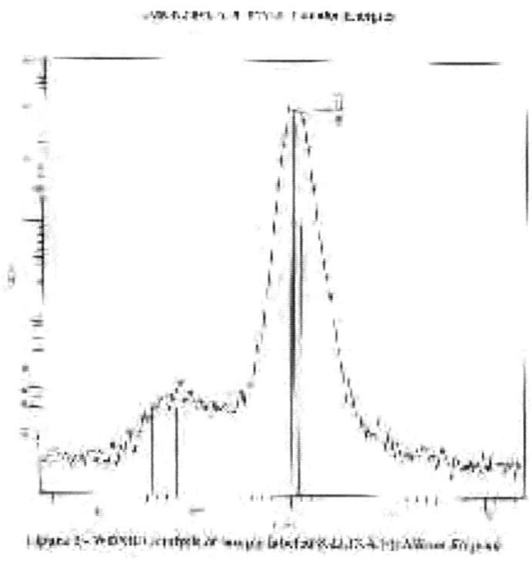

Figure 19: *A view of one of the analyses [116]-[118] signed by laboratory directors showing the peack of the Silicon synthesized from Carbon and Oxygen.*

the ICNF

$$O(8,16,0^+,17.9991) + O(8.16,0^+,17.9991) \rightarrow$$
$$\rightarrow S(16,32,0^+,31.9720) + \Delta E, \quad (100)$$

where

$$\Delta E = 4.0262\ amu = 3,7483,3922\ MeV =$$
$$= 5.6950 \times 10^{-13} BTU\ per\ synthesis, \quad (101)$$

and other ICNF that were announced in video [113] and paper [114] of 2011 (see also lecture [115]) following systematic analyses [116] - [120] also signed by the laboratory director (see the sample analysis of Figure 19).

It should be stressed that the above tests were primarily done to establish the *existence* and *control* of ICNF in parts per millions under the expectation that such an evidence was sufficient for funding the construction of industrial hadronic reactors, by keeping in mind that the efficiency of the reactors was limited by the low operating pressure, the low power of the DC generator and other reasons.

The rather voluminous amount of information collected on ICNF indicates that the achievement of a an *industrial* hadronic reactor capable of producing truly controlled and radiation free nuclear energy is indeed within current technological reach via the use of: a specially built, $100\ KW$ DC power unit of the particular type used for the synthesis of the neutron (Section 8.2.3); the use of bigger pressures; and other engineering means. To see it, recall that the Nitrogen synthesis of Ref. [103] produces $1.57 \times 10^{-15} BTU$ *per fusion.* The plasma surrounding a '1 cm DC arc contains about 10^{30} atoms. The extrapolation of available data indicates the realistic possibility of producing of 10^{21} fusions *per hour* with the ensuing delivery of $10^6\ BTU/h$ which are fully sufficient to generate the needed electric power plus the delivery of clean energy of nuclear origin. Alternatively, as shown in video [113], the third hadronic reactor for the synthesis of the Silicon was already producing electricity when operated at $1,000\ psi$ pressure under $50\ KW$ power. The production of free, clean, nuclear energy when operated at $5,000\ psi$ pressure with $100\ KW$ DC power is rather plausible.

A study of ICNF presented at the Teleconference is available in Ref. [219] and in the lecture recorded by I. B. Das Sarma [0].

It is regrettable for the environment that billions of dollars of public funds have been and continue to be invested in conventional nuclear fusions despite known insufficiencies of quantum mechanics, while it has not been possible to secure until now comparatively small funds for the test of the indicated ICNF.

8.2.9-III. HyperFusions of natural nuclei and pseudo-nuclei. The biggest problem that has prevented the achievement to

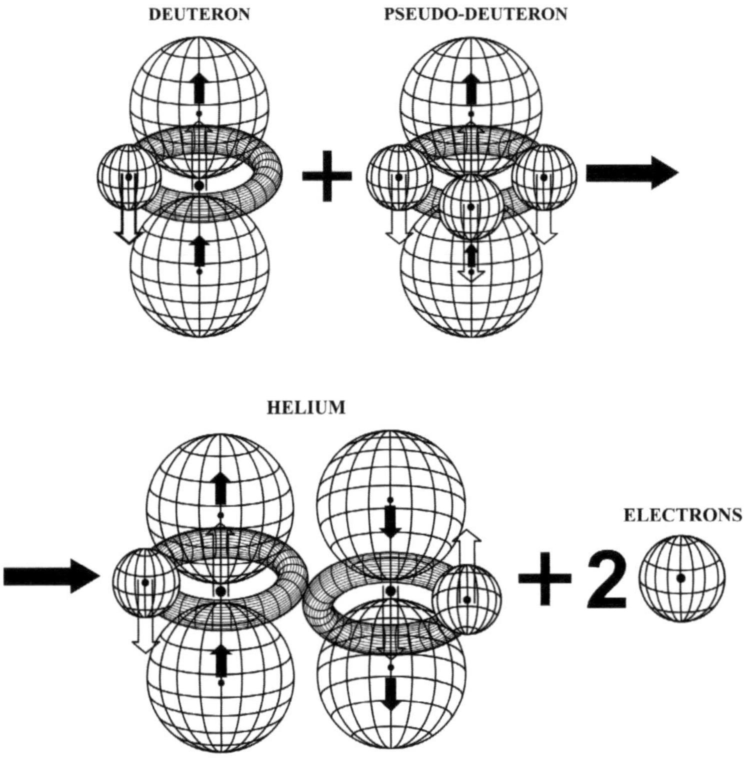

Figure 20: *A serious environmental problem facing mankind is the lack of achievement in about a century of clean cointrolled nuclear fusions because of the extreme value of the Coulomb repulsion between nuclei at* 1 fm *mutual distances (the Coulomb barrier). In this figure, we illustrate the new HyperFusion which has been conceived [214] to eliminate the Coulomb barrier via the synthesis of the negatively charged pseudo-Deuterons* $D(-1, 2, 1)$ *(Section 8.2.8) which are then attracted by natural Deuterons* $D(1, 2, 1)$ *in the appropriate singlet couplings to synthesize the Helium* $D(-1, 2, 1_{\uparrow}) + D(1, 2, 1_{\downarrow}) = He(2, 4, 0) + 2e$ *Note the impossibility for the indicated HyperFusion to produce neutrons or other harmful radiation (see Section 8.2.9-III for details).*

date of controlled nuclear fusions is the *Coulomb barrier,* namely, the extremely big *repulsive Coulomb force* between natural nuclei caused by their positive charge which, from Eq. (6), has the value of hundreds of Newtons.

In this section, we report the research done to date on a basically new type of nuclear fusion, called *HyperFusion,* first proposed in Section 2.7.3 of Ref. [214], which occurs between natural, positively charged nuclei and synthesized negatively charged nuclei (Section 8.2.8), by therefore turning the repulsive Coulomb force into an attraction.

The principle of HyperFusion is elementary. Recall from Section 8.2.2 that at 10^{-13} cm the electron and the proton experience the extremely big *attractive* Coulomb force of 230 Newtons. Nevertheless, quantum mechanics admits no bound state between the electron and the proton at short distance, by therefore confirming Einstein's view on the lack of completeness of quantum mechanics beyond scientific doubts [1]. The completion of quantum into hadronic mechanics then allowed the identification of all engineering needs for the synthesis of the electron and the proton into an unstable particle with 900 seconds mean life, the neutron.

The principle of the new HyperFusion is essentially the same. At 10^{13} cm mutual distance, electrons are attracted by nuclei with a Coulomb force of hundreds of Newtons. The technology established for the Rutherford compression of the electron inside the proton is also applicable to the compression of electrons inside nuclei, resulting in the synthesis of negatively-charged nuclei, called *pseudo-nuclei* [214] that are evidently unstable, but possess nevertheless mean lives of the order of seconds, thus being usable for industrial applications. HyperFusion is then completed by established technologies for the separation of pseudo-nuclei from their originating plasma and their exposure to natural nuclei.

For the case of the negatively charged pseudo-Deuterons obtained from the DNS filled up with a Deuterium gas (Section 8.2.) [214],

$$D(1,2,1) + (e^-_\downarrow, e^-_\uparrow) \rightarrow \hat{D}(-1,2,1), \tag{102}$$

we have the predicted HyperFusion [214]

$$D(1, 2, 1^+_\uparrow, 2.0141\ amu) + \hat{D}(-1, 2, 1^-_\downarrow, 2.0141\ amu) \rightarrow$$

$$\rightarrow He(2, 4, 0, 4.0026\ amu) + 2\beta^- + \Delta E, \tag{103}$$

$$\Delta E = 0.0222 amu = 20.67\ MeV. \tag{104}$$

in which the opposite alignment of spins (which is necessary for the conservation if the angular momentum) occurs naturally due to the opposing charges and magnetic moments (Figure 20).

Note that the above fusion not only eliminates the *repulsive* Coulomb force between natural nuclei, but turns the repulsion into an *attraction* due to the opposite charges, as a result of which Deuterons and pseudo-Deuterons are naturally attracted to the mutual distance of $10^{-13}\ cm$ at which nuclear forces are activated and the fusion becomes inevitable.

Recall also that a controlled quantum mechanical fusion of two Deuterons into the Helium has been essentially impossible due to the need for a *controlled* singlet coupling of the two Deuterons which is needed for the conservation of the angular momentum (see the hadronic laws for nuclear fusions in Ref. [102]). By comparison, the coupling of a Deuteron and a pseudo-Deuteron is naturally singlet as expressed in reaction (103) due to their opposite charges and magnetic moments, by therefore providing a second facilitation for a controlled nuclear fusion over conventional attempts.

Note also that the fusion $D + \hat{D} \rightarrow He + 2\beta$ can be initiated and stopped on demand, and can be controlled via the control of: the pressure of the Deuterium gas; the energy of the DC arc; its voltage; the flow of the Deuterium gas through the arc; and other engineering means.

Note finally that the interactions with representation (45) bonding together the two Deuterons into the Helium constitute a third explicit form of *fifth interactions* according to its definition given in Section 7.2.3.

8.2.10. The problem of recycling nuclear waste. As it is well known, automotive production is under reorganization in the U.S.A. and in other developed countries to replace fossil fuels operated cars with electrically operated cars. This reorganization implies a consequential, not sufficiently spoken, exponential increase of electricity produced from nuclear power plants and other sources. As a consequence, the problems of recycling highly radioactive nuclear waste and the achievement of clean combustion of fossil fuels are becoming some of the most serious environmental problems facing mankind today.

In this section we shall indicate the implications of the EPR argument for the recycling of nuclear waste, while the achievement of full combustion of fossil fuels is treated in the next section.

Due to various oppositions, it has not been possible to store radioactive nuclear waste in depositories such as that in the Yucca Mountain in the U.S.A. Consequently, nuclear waste is stored nowadays by the nuclear power plants themselves in their own facilities. Such a storage has already reached safety limits with an expected unreassuring surpassing of safety limit under the ongoing automotive reorganization from fossil fuels to electric cars.

Also, all attempts at an effective recycling of nuclear waste via quantum mechanical technologies have failed to date. Therefore, a most unreassuring aspect of the current condition is the lack of a visible search for *new* forms of waste recycling, despite the availability of large corporate and governmental funds, and the continuation of the century old oblivion of the EPR argument [1].

By recalling that any transportation of radioactive nuclear waste is prohibited by popular opposition, the indicated unreassuring condition mandates at least the *search* by responsible societies of the only viable solution, that of developing new technologies for the recycling of nuclear waste by the nuclear power plants themselves in their own facilities. Among all possible solutions, the most desirable recycling is that via the

Figure 21: *A second serious environmental problem facing mankind is the recycling of radioactive nuclear waste now stored by nuclear power plants in their facilities. It is unfortunate for mankind that the century old opposition to the the EPR argument by mainstream physics generally discredits the search for stimulated recycling of nuclear waste because not permitted by quantum mechanics.*

stimulated decay of radioactive waste in such a manner of reducing mean lives from thousands of years down to days or minutes.

It is known that such a recycling is impossible for quantum mechanics because the alteration of the mean-life of radioactive nuclei would imply a violation of the Poincaré symmetry which is at the foundation of special relativity.

In this section, we outline the following three recycling of nuclear waste proposed in Ref. [97], Section 4-III-A, page 342 on, which are made possible by the completion of quantum into hadronic mechanics:

8.2.10-I. Recycling of radioactive hospital waste. Tests conducted in 2011 at the U. S. publicly traded company Magnegas Corporation have indicated that a 300 KW PlasmaArcFlow Hadronic Reactor (Figure 21) can apparently recycle lightly radioactive liquid waste from hospitals by triggering their decays via the deformation of their nuclei when exposed to intense electric and magnetic fields into such a prolate

Figure 22: *In this figure, we show a 300 Kw PlasmaArcFlow Recycler produced and sold by Magnegas Corporation (now Taronis Corporation) which is recommended for the recycling of radioactive hospital waste, since the extreme electric and magnetic fields of the submerged electric arc predict their stimulated decay.*

shape to cause the disintegration of naturally unstable nuclei due to internal Coulomb repulsion between aggregated protons at the extremities.

Note that such a deformation is impossible for quantum mechanics due to the representation of nuclei as massive points. By contrast, said deformations are fully possible for hadronic mechanics due to the representation of nuclei as extended and, therefore, deformable according to Eqs, (45)-(46) for which the stimulated decay of nuclear waste is reduced to the intensity of the electric and magnetic fields.

Jointly, the indicated 300 KW Plasma-Arc-Flow hadronic reactor sterilizes the liquid waste due to its exposure to the high temperature of the arc.

8.2.10-II. Recycling nuclear waste via stimulated neutron decays. A primary reason for suggesting the rather inexpensive test of the stimulated neutron decay in Section 8.2.9-I is that, in the event successful, it would allow an effective recycling of radioactive nuclear waste via the stimulated decay of some of its peripheral neutrons. This possibility is illustrated in Ref. [97], page 340, with the predicted stimulated decay of $42 - Mo - 100$, which is an unstable nuclide with mean -life of 10^{19} years. Yet, a stimulated neutron decay would imply the transmutation of $42 - Mo - 100$ into $43 - Tc - 100$ according to the reaction

$$\hat{\gamma}(0, 0, 1, 1.295\ MeV) + Mo(42, 100, 0, 99.9074\ amu) \rightarrow$$

$$\rightarrow Tc(43, 100, 1, 99.9076 amu) + \hat{e}^{-}(-1, 0, 0, 1.295\ MeV), \tag{105}$$

by turning the mean life of 10^{19} $years$ of $42 - Mo - 100$ into the mean-life of18.5 $seconds$ for $43 - Tc - 100$, with natural decay

$$Tc(43, 100, 1, 99.9076\ amu) \rightarrow$$

$$\rightarrow Ru(44, 100, 0, 99.9042\ amu) + e^{-} + \nu\ (or\ a), \tag{106}$$

where one should note that the final nuclide is stable and there is no emission of neutronic or other damaging radiation.

From data [99], we know that the indicated transmutation produces heat for $3.475\ MeV$ per reaction plus electricity generated by easily trappable electrons.

For brevity, interested readers may inspect Ref. [97], page 342 for a possible industrial recycling of nuclear waste via a beam of resonating photons causing two or more stimulated neutron decay per nucleus.

8.2.10-III. Recycling nuclear waste via pseudo-proton irradiation. The 2020 Teleconference [0] included discussions on the possible recycling of radioactive nuclear waste via their irradiation with pseudo-protons (Section 8.2.7). In this case, unlike stimulated decay (105)-(105), we have the predicted reaction via the use of two pseudo-protoids with spin 0

$$2\bar{p}^{-}(-1, 1, 01.0073\ amu) + Mo(42, 100, 0, 99.9074\ amu) \rightarrow$$

$$\rightarrow Zr(40, 102, 0, 101.9229\, amu)n + \Delta E, \tag{107}$$

in which case the 10^{19} *years* main-life of $40 - Zr - 102$ is reduced to the of 2.9 *seconds* mean-life of $40 - Zr - 102$.

Since the corporate handling of nuclear waste is prohibited by law, systematic research on the suggested stimulated recycling of radioactive nuclear waste by nuclear plants themselves is left to governmental laboratories of countries interested in developing new technologies permitted Einstein's legacy.

In closing this section, it should be indicated that mainstream physicists generally attempt to discredit the search for the new clean nuclear energies on grounds that they are not predicted by quantum mechanics, without any repetition of the numerous inexpensive tests done to date, by ignoring the gross insufficiencies of quantum mechanics in nuclear physics (Section 3.2), and in oblivion of Einstein's view that "quantum mechanics is not a complete theory" (Figure 20).

Due to its importance for society, the problem of the recycling of nuclear waste was addressed in the opening lecture of the 2020 Teleconference by J. Dunning-Davies (see the recorded lecture [0]) and Ref. [220].

8.3. Applications of EPR completions in chemistry.

8.3.1. Representation of valence electron bonds. There is no doubt that, beginning with the 1916 pioneering contribution by G. N. Lewis, followed by numerous advances, the 20th century valence bond theory has allowed historical discoveries in chemistry.

Yet, it is well known that science at large, and chemistry in particular, will never admit final theories. Advances generally depend on the identification of open problems followed by due scientific process in the proposed solutions.

One of the best kept secrets of the best Ph. D. schools in chemistry is that there is no possibility for quantum mechanics and, therefore, for quantum chemistry, to represent the *attraction* between the *identical* electrons in valence bonds because the Schrödinger equation for an electron pair, Eq. (1), can only predict the extremely big *repulsion*, of 230 Newtons at the 10^{-13} cm mutual distance of a valence electron bond (Eq. (6) and Section 3.5).

A truly *attractive* force between identical valence electron pair was achieved in 2001 (see Chapter 4 of Ref. [30]) inspired by the last statement of the 1935 paper by Einstein, Podolsky and Rosen [1] according to which the *wavefunction* of quantum mechanics and chemistry cannot describe all "elements of reality."

As reviewed in Section 7.4, the much needed attractive force was achieved thanks to the completion of quantum into hadronic chemistry, resulting in a Hulten-type new bond, of type (61) called *strong isovalence bond*, which is so strongly attractive to generate a quasi-particle called *IsoElectronium* (IE)

$$IE = (\hat{e}_{\downarrow}^{-}, \hat{e}_{\uparrow}^{-})_{hm}, \tag{108}$$

with double elementary charge, null spin and null magnetic moment.

In this section, we hope to indicate the unreassuring implications for our environment of molecular models without a clear attraction between valence electron pairs (Figure 18).

8.3.2. Representation of molecular data. Historical advances have been made during the 20th century in the collection and representation of molecular data. Yet, a belief in the achievement of final knowledge would imply the existing from the boundaries of science since so much remains to be understood.

Quantum mechanics allowed the representation of the experimental data of the Hydrogen atom from first axiomatic principles with an incredible accuracy. By comparison, when two Hydrogen atoms are bonded together in the Hydrogen molecule, quantum chemistry misses 2% of the

Figure 23: *The third serious environmental problem facing mankind is the lack of full combustion of fossil fuels. It is unfortunate for our environment that mainstream chemistry generally discredits its solution via the EPR completion of quantum chemistry in oblivion of the inability by quantum chemistry to represent the attraction between valence electron pairs and Einstein's legacy that quantum wavefunctions cannot represent all "elements of reality."*

molecular binding energy from first axiomatic principles without adulterations, which percentage is far from being ignorable because it corresponds to about $950\ BTU$.

The inability by quantum chemistry to achieve an exact representation of the binding energy of the Hydrogen and other molecules is clear evidence of the appearance in valence electron bonds of short range nonlinear, non-local and non-potential, thus non-Hamiltonian interactions typical of the EPR entanglement (Section 7.2.3) that are absent at the large mutual distances of the atomic structure.

In fact, the EPR completion of quantum into hadronic chemistry, and the achievement of a strong valence bond, have allowed the following advances:

1) The representation of the binding energies of the Hydrogen [31] and water [32] molecules which is *exact to the desired decimal value.*

2) Perturbative calculations used in the above results that have a con-

vergence at least one thousand times faster than the convergence of the corresponding quantum chemical calculations, thanks to the very low value of the isotopic element $\hat{T}$ inserted in all associative products $A\hat{\times}B = A\hat{T}B$ (Section 7.5).

3) The representation of the diamagnetic character of the Hydrogen molecule which is not possible for 20th century weak valence bonds.

4) The explanation of the reason why the Hydrogen molecule can only accommodate *two* Hydrogen atoms in its stable configuration, resulting in a *restricted three-body model of the hydrogen molecule,* with the consequential, first known admission of analytic solutions that have evident importance for the environment, e.g., to achieve full combustion of fossil fuels.

5) Consequential advances in other molecules, such as the *iso-Helium model* [122] and other advances [123].

8.3.3. The new chemical species of magnecules. Nowadays, we release in our atmosphere about *35 billion tons of contaminants per year* mostly composed by Carbon Monoxide $CO = C - O$ (where $-$ represent valence bond) and HydroCarbons HC. As it is well known, this environmental problem is caused by the incomplete combustion of fossil fuels (Figure 23). By noting that CO and HC are themselves combustible, the indicated environmental problem is caused by the lack of complete combustion of fossil fuels. In turn, this is due to a combustion insufficient for the dissociation of the valence bonds of fuels such as gasoline or diesel. Note that the dissociation of fuel molecules into their atomic constituents is a necessary pre-requisite for the chemical synthesis of CO and HC.

In view of the above unreassuring data, a considerable scientific and industrial effort was done in early 2000 to conceive, test and produce *fuels with the new chemical species of magnecules* (see Chapter 8 of Ref. [30] and the U. S. patent [123]). The new species of magnecules (to distinguish them from molecules) has essentially the same *atomic* components of fossil fuels and it is stable in pressure thanks at ambient temperatures, yet *the binding energy of magnecules is much smaller than that of molecules* as a necessary condition to achieve full combustion, namely, a combustion without detectable CO and MHC in the exhaust.

The new species of magnecules is essentially composed by clusters of atoms (such as H, C, etc.), dimers (such as $H - O$, $C - H$, etc.), and ordinary molecules (such as $H - H$, $C - O$, etc.) bonded together by the attractive force called *magnecular bond* between opposing magnetic polarities of toroidal configurations of the orbits of peripheral electrons, which magnecular bond is stable under ambient temperature.

As an illustration, by denoting the magnecular bond with the symbol "$\times$", samples of *elementary magnecules* are given by (see Figure 24)

$$H \times H, \ \ O \times O, \ \ C \times O, etc. \tag{109}$$

By noting that electron orbitals can be controlled under very big electric and magnetic fields [30], new fuels with magnecular structure are produced via specially designed hadronic reactors converting fossil fuels into their gaseous magnecular form via a submerged electric arc. Water is added to fossil fuels to increase the content of Hydrogen and Oxygen of the magnecular species by therefore improving its environmental qualities [124].

Stock cars produced to run on compressed natural gas, but operated on compressed magnegas release an exhaust containing no appreciable Carbon Monoxide CO or HydroCarbon HC; and Oxygen O_2 up to 14% (see Figure 22 and the documentation presented at the 2016 International Summit on the Environment, Hainan Island, China [126]).

In view of the importance of the above features for the environment, the publicly traded company *Magnegas Corporation* was organized for the production and sale world wide of the gasification of fossil fuels into the environmentally friendly *magnegas* (chemical symbol MG, Figure 25)

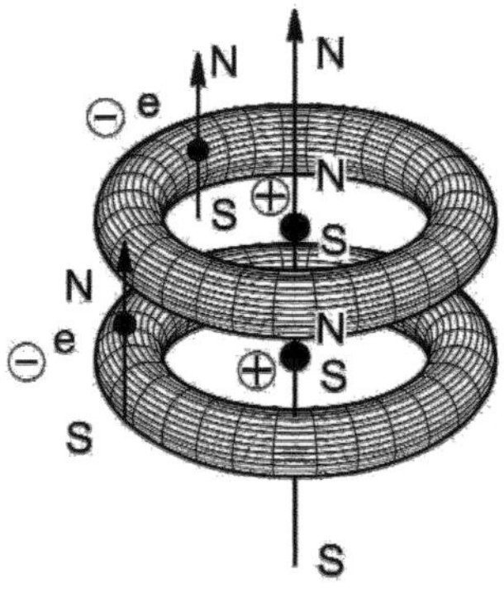

Figure 24: *An illustration of the elementary magnecule such as $H \times H$, $O \times O$, $C \times O$, etc., where one can see: the toroidal polarization of atomic orbitals; the attraction between opposing magnetic polarities; and the exposure of nuclei out of their electron cloud with the proper spin coupling, which features are essential for the ICNF of Section 9.2.9-II.*

Figure 25: *A view in the left of a Ford Cavalier produced to run on compressed natural gas and run on compressed magnegas with no appreciable CO and HC in the exhaust, and a view in the right of a Ferrari 308 GTS 1981 converted to run on compressed magnegas and shown at the Moroso International Race Track, Florida, to be competitive with same Ferraris running on gasoline.*

[125].

It should be indicated that *the author does not recommend the large scale use of Hydrogen as an alternative automotive fuel to gasoline* because: 1) Cars running on compressed Hydrogen are expected to produce neutrons (Section 8.2.3); 2) The large scale use of Hydrogen would cause a prohibitive Oxygen depletion; 3) Hydrogen seeps through walls and immediately raises to the Ozone Layer with ensuing rapid chemical reaction $H_2 + O_3 \rightarrow O_2 + H_2O$; and for other environmental problems. [30].

8.3.4. Experimental verifications of the new species of magnecules. The chemical composition of gases is nowadays analyzed via Gas Chromatographers Mass Spectrometers (GC-MS), InfraRed Detectors (IRD) and other equipment all designed, specifically, to detect *molecules,* that is, atoms bonded together by valence bonds with ensuing large binding energy. In said instruments, molecules are first exposed to an ionization beam whose energy is below the molecular dissociation energy; the molecules are then processed by a reactant in the column; and then eluded with a speed inversely proportional to their masses for identification.

By noting that magnecules have a binding energy which is about 10% that of molecules, GS-MS are basically unable to detect magnecules because their ionization beams destroy the very species to be detected, since they terminate all magnecular bonds and reduce the species to conventional molecules.

In view of the above, magnecules can only be indirectly detected via GC-MS/IRD and the findings confirmed via direct measurements of the characteristics of the magnecular gas, such as flame temperature, BTU content, etc. [126].

The instruments used so far for the analysis of fuels with magnecular structure are GC-MS equipped with IRD (GC-MS/IRD) so as to test first the gas un the GC-MS used with special provisions, such as the lowest possible ionization energy, the lowest possible column temperature, and the longest possible elusion time so as to minimize the destruction of the species to be detected. The clusters identified in the GC-MS are then tested via the IRD *without transferring the gas to a separate IRD because of the impossibility to combine with certainty results from different tests.* Magnecules are detected when the clusters identified in the GC-MS have no IR signature because lack of IR signature implies the lack if existence of internal molecular bonds.

The GC-MS/IRD tests that were originally done by the author in 1998 are reported in Chapters 8 and 9 of Ref.[30], Section 8.7 on). More recent independent experimental verifications via GC- MS/IRD are available in Refs. [127] [128]. Comparison of the original tests [30] with then more recent tests [127] [128] shows that the detection of magnecular clusters is decreased in the latter tests due to the increase of the ionization energy of the GC-MS which was confirmed by comparison of data in the relative manuals.

The best evidence establishing the existence of a *new* chemical species is given by direct measurements of the characteristics of magnegas since said characteristics cannot be explained with quantum chemistry.

The first of these anomalous characteristics is given by extensive tests conducted by scientists of the *Institute for Ultra Fast Spectroscopy and Laser of the City College of New York,* [129] [130] which established that *magnegas synthesized from fossil fuels has a flame temperature in air of* $10,597\ F = 3,647\ C$ which is 56% *bigger* than the hydrogen flame temperature in air $2,045\ C$. The increased flame temperature explains the absence of CO and HC in magnegas exhaust due to their combustion. Additionally, the increased flame temperature makes magnegas a fuel particularly suited for steel mills, refineries and other applications [124]-[126].

An intriguing feature of magnegas is its energy content because *magnegas cuts a 12" thick steel slab faster than acetylene* (C_2H_2) *which pos-*

sesses $1,498\ BTU/cf$ [132]. This anomaly property is established by the fact that a conventional chemical analysis of magnegas done at maximal ionization energy and column temperature reveals that magnegas is composed by

$$MG:\ 54\%\ H_2,\ \ 31\%\ CO,\ \ 15\%\ HC, \tag{110}$$

with corresponding energy content of $325,\ 323\ ,1,500\ BTU/cf$. Hence, according to quantum chemistry, magnegas synthesized from petroleum should contain a maximum of

$$54\%\ 325 + 31\%\ 323 + 15\%\ 1,500\ BTU/cf = 431\ BTU/cf, \tag{111}$$

namely, an amount basically insufficient to cut any metal, let alone 12" thick steel plates, thus confirming that *magnegas cannot be quantitatively described via quantum chemistry.*

The description of the energy content of magnegas according to hadronic chemistry is essentially the following. Let us denote the molecular bond with the symbol "$-$" and the magnecular bond with "$\times$." It is generally believed that the combustion of Hydrogen and Oxygen according to the reaction

$$H - H + (1/2)O - O \rightarrow H - O - H + 57\ BTU/cf. \tag{112}$$

However, the two Hydrogen atoms are separated in the $H - O - H$ molecule. Therefore, in order to verify the principle of conservation of the energy, the combustion of Hydrogen and Oxygen must produce $104\ kcal/mole$ for the $H - H$ dissociation, plus $45\ kcal/mole$ for the $O - O$ dissociation, plus $57\ kcal/mole$ of produced heat, for a total of $206\ kcal/mole$. In the event the Hydrogen and Oxygen species have a magnecular structure $H \times H$ and $O \times O$, the binding energy is only 10% of the molecular value. In this case, the combustion of the *magnecular* species $H \times H$ and $O \times O$ produces the total energy of $97 + 40 + 57\ kcal/mole = 194\ kcal/mole$

$$H \times H + (1/2)O \times O \rightarrow H - O - H + 194\ kcal/mole, \tag{113}$$

which is 3.4-times the energy produced by the molecular species, Eq. (113), resulting in the value of $1,465\ BTU/cf$ which is comparable to the energy content of acetylene $1,498\ BTU/cf$, by providing in this way the only known quantitative explanation of the reason magnegas cuts faster than acetylene under essentially the same atomic structure.

In conclusion, *the energy content of molecular gases is constant while, by comparison, the energy content of magnecular gases is variable because it depends on the original liquid feedstock, the power of the hadronic reactor, and other engineering data.*

8.3.5. Magne-Hydrogen and Magne-Oxygen. The anomalous temperature and energy content of magnegas is due to the anomalous character of its primary components known as *Magne-Hydrogen* (symbol MH), *Magne-Oxygen* (MO) and *Magne-Carbon-Monoxide* (MCO) [133]. The indicated new species have been separated from magnegas via Pressure Swing Adsorption (PSA) equipment, also called molecular seeving, and their anomalous specific weight has been confirmed by independent measurements [134]-[137]. By recalling that MH constitutes more than 50% of MG, the anomalous energy content of MH is established by that of MG. Tests on the increase of the efficiency of fuel cells via the use of MH and Oxygen are reported in Ref. [124]. The case of a 3-atomic magnecule or molecule is indicated in Figure 26.

An important experimental evidence in the above tests is that the specific weight of the magnecular species MH, MO and MCO increases with the number of times the separated gas is passed again through the PSA equipment. This is a clear indication of the effect known as *magnecular accretion*, namely, the increase of the mass of magnecule with the increase of pressure and other treatments.

In turn, the above effect indicates that *the Avogadro number is not expected to be a constant for magnecular gases* [30], which feature is suggested for test by interested chemists jointly with other innovative mea-

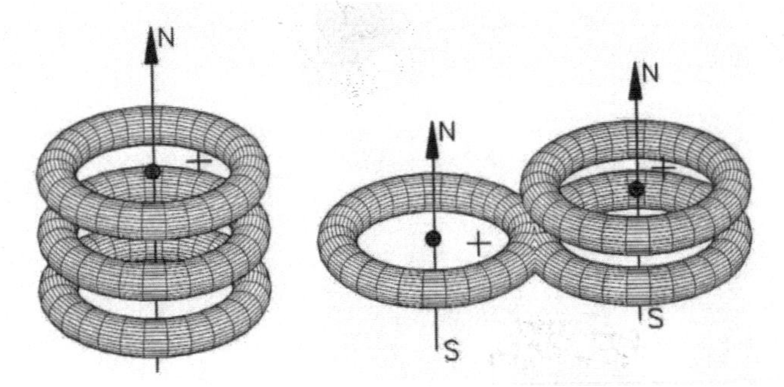

Figure 26: *An illustration of the magnecule* $3H = H \times H \times H$ *that should be compared with the molecule* $H_3 = H - H - H$. *The main difference is that the latter is unstable, while the former is stable at ambient temperature according to repeated detections of 3 amu in the same* MH *gas. Such a difference is due to the independence of the magnecular bond from the number of bonded atoms compared to the lack of existence of a valence electron triplet.*

surements.

An understanding of the above intriguing feature is the following. Recall from Section 3.5 that another insufficiency of quantum chemistry is the impossibility of achieving an *attractive* force between the water molecules if the liquid state due to their diamagnetic and dielectric characters. It appears that such an attractive force is of magnecular type since the Hydrogen atoms H in the water molecule $H - O - H$ has a toroid polarization of its orbit for the proper bonding to the corresponding valence electron of the Oxygen, thus allowing a magnecular attraction of the type [138]

$$H - O - H_{\downarrow} \times H^{\uparrow} - O - H^{\uparrow} \times H_{\downarrow} - O - .... \tag{114}$$

According to this view, the magnecules of the new species MH, MO and MCO are *quasi-liquid* with consequential magnecular accretion and the predicted lack of constancy of the Avogadro number.

Chemical analyses have repeatedly established that MH is composed by atomic masses from 1 amu to hundreds of amu, thus suggesting a structure of the type

$$MH: \ \{H, \ H-H, \ H \times H, \ H \times H, \\ H-H \times H, \ H-H \times H, ...\}, \quad (115)$$

with a corresponding structure for MO and MCO.

In view of its special features, including the increased specific weight, increased energy content and predicted increase of the liquefaction temperature, MH is expected to be particularly significant for the aerospace industry as well as for gasoline refineries, fertilize production and other fields. MO is expected to be signifiant for medical applications, e.g., for use in ventilators for persons affected by the Corona Virus because its magnecular structure is expected to decompose at contact with lungs, with ensuing release of sterilizing UV radiation as it is the case for the decomposition of Ozone. MCO has been conceived and tested for the primary aim of developing the HyperCombustion indicated (Section 8.3.7).

8.3.6. The HHO combustible gas. One of the most intriguing fuels with magnecular structure is the *HHO gas* [139] which is produced in the gasification of water via a specially designed water electrolyzer developed by the U. S. company *Hydrogen Technologies Applications, Inc.*

Some of the unique features of the HHO gas, which are manifestly outside any serious representational capability by quantum chemistry, are:

1) The HHO gas instantly cuts bricks, tungsten and other hard material at content in air or under water;

2) The combustion of the HHO gas occurs without any need for atmospheric Oxygen, since HHO is composed by a stochiometric ratio of Hydrogen and Oxygen; and

3) The combustion exhaust of the HHO gas are composed by water vapor without any contaminant.

By using a cautious scientific language, Ref. [139] (see also Ref. [124]) presents a series of *measurements* on the HHO gas conducted with various instruments and their tentative representation with the new chemical species of magnecules.

It should be indicated that all hadronic reactors indicated in this section have a *negative* energy balance in the sense that they require energy to produce a gaseous, environmentally friendly fuel, unless, jointly with the gasification of the liquid feedstock, the hadronic reactors synthesize new elements, such as the synthesis of $Si - 28$ from $C - 12$ and $O - 16$, in which case the energy balance is positive even for nuclear synthesis in ppm.

8.3.7. Hyper-Combustion. At the dawn of the third millennium, the combustion of fossil fuels is essentially the same as it was some fifty thousands years ago, because in our civilian, industrial and military engines we essentially strike a spark and lit the fuel.

This section is intended for scientists interested in the search for a basically new form of combustion as a necessary condition for the future achievement of a sustainable life of our planet.

Recall from Section 8.3.3 that magnegas does achieve *full combustion,* namely a combustion without appreciable combustible contaminants in the exhaust.

Based on such a result, a main drive of the physical and chemical studies reported in this section has been the achievement of full combustion, this time, for fossil fuels as commercially available, including gasoline, diesel, methane, acetylene, et al.

Recall that our combustion of about 36 *billion* barrels of crude oil per year releases in the environment about 10 billion barrels per year of contaminants such as CO and HC. However, it is important to note that

Figure 27: *A view of the equipment developed by Hadronic Technologies Corporation to test Hyper-Combustion in a four cylinder electric generator showing (from bottom right) Control Module, Variac and four ICNF Activation Banks (Section 3.3.7).*

CO and HC are themselves combustible. This establishing that *the temperature of combustion in our civilian, industrial and military engines is insufficient to achieve full combustion.*

The studies reported in this section have led to the formulation of a new form of combustion known as *Hyper-Combustion* [140] under development by the privately held *Hadronic Technologies Corporation* which can be defined as follows: *the hyper-combustion of Carbon with atmospheric Oxygen comprises the conventional chemical combustion plus a controllable number of engineering means causing the synthesis of Silicon and other light, natural and stable elements without the emission of harmful radiation and without the release of radioactive waste.*

The engineering realization of the hyper-combustion is based on the use of a specially designed DC power unit delivering an arc that: 1) Ionizes the fuel; 2) Creates magnecules $C \times O$ between toroidally polarized stable isotopes of Carbon and Oxygen; and 3) Triggers Intermediate Controlled Nuclear Fusions (ICNF, Section 8.2.9) in parts per million (ppm) of the indicated magnecule Carbon and Oxygen into the Silicon.

Recall from Nuclear Data [99] that Carbon has the following two stable isotopes with relative abundance $6-C-12$, 98.898%, $6-C-13$, 1.11% and that the Oxygen has the following three stable isotopes with relative abundance $8-O-16$, 99.76%, $8-O-17$, 0.037%, $8-O-18$, 0.200%. Consequently, when dealing with nuclear fusions in ppm, the significant isotopes are $6-C-12$ and $8-O-16$ with related nuclear fusion into the $14-Si-28$ studied in Section 9.2.9-II, Eqs. (98)-(99). This fusion has been experimentally verified with the Third Hadronic Reactor (Figures 15 and 16, video [113], sound [108], and verifications [114]-[120]).

Note that the aim of hyper-combustion *is not* that of replacing fossil fuels with controlled nuclear fusions, but that of using Carbon and Oxygen as fuels for nuclear fusions mainly intended to maximize the energy output of crude oil and achieve full combustion without harmful radiations or radioactive waste.

To minimize possible misrepresentations, it should be reported that Hadronic Technologies Corporation has manufactured an equipment (Figure 27) testing Hyper-Combustion in a four cylinders electric generator with tungsten tip spark (patent pending) which comprises: 1) A control module to set maximal power and arc settings per selected engine; 2) A variac to operate with minimal and maximal power; and 3) Four ICNF Activation Banks, one per cylinder. The equipment produced the desired increase of the combustion temperature at which CO, HC and other combustible contaminants burn by producing a first increase of power output. The same equipment has produced a second increase of energy output due to the activation of ICNF in ppm of Carbon and Oxygen into Silicon within the thermodynamical limit of the electric generator.

In closing this section, it should be reported that mainstream chemists generally dismiss the *anomalous characteristics* of magnegas as being due to *anomalous measurements,* without repeating the measurements and without considering the fact that said measurements were done by the highly authoritative Institute for Ultra Fast Spectroscopy and Laser of the City College of New York [130] [131].

Additionally, the same chemists generally dismiss the existence of the new species of magnecules on grounds that the identified *anomalous clusters* can be reduced by the GC-MS to their *molecular* constituents, by ignoring that: 1) Stable clusters systematically detected by GC-MS must have an internal atomic or molecular bond to exist; 2) The ionization energy used in the decomposition of the clusters to their molecular constituents destroys the very species to be detected; and 3) The dismissal is done in oblivion of Einstein's view on the limitations of quantum mechanics and chemistry, as well a without the proof in refereed journals that quantum chemistry provides a quantitative representation of magnegas anomalous features.

Objections against magnegas and magnecules voiced in social media, rather than via publications in refereed journals, has significantly undermined and delayed the solution to our environmental problems. In fact, as it is well known, social media plays in the hands of those who profit by shorting the stock of publicly traded companies, by therefore causing significant losses to investors eager to support the development of new environmental technologies for a sustainable future of our planet.

8.4. Applications of EPR completions to special relativity.

8.4.1. Foreword.

In this section, we show that the methods developed for the proof of the EPR argument [1] imply *a geometric unification of Einstein's special and general relativities for the exterior problem of point-like particles in vacuum, as well as their extension for interior dynamical problems of extended particles within physical media* .

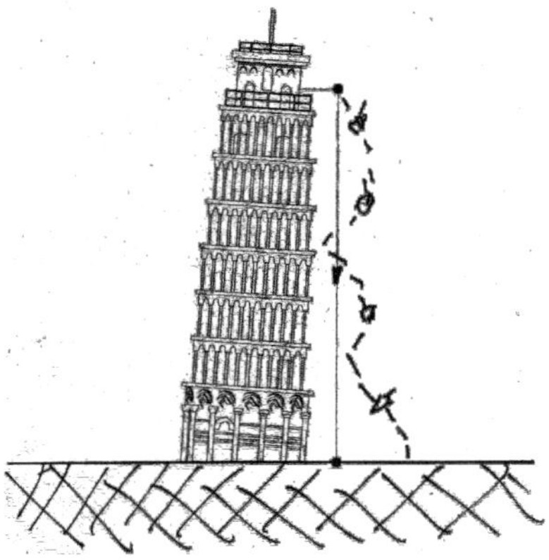

Figure 28: *A schematic view of the experiments done by Galileo Galilei at the end of the 16th century to measure the acceleration due to gravity by dropping from the top of the Pisa tower balls with different masses assumed to be point-like, thus ignoring the resistance due to our atmosphere (vertical line). The axiom - preserving iso-Galilean isorelativity aims at the dynamics of extended bodies, thus including atmospheric resistance (wiggly line).*

8.4.2. Applications of the EPR completion to Galileo's relativity. As it is well known, the Galileo symmetry $G(3.1)$ and relativity are exactly valid for the description of non-relativistic conservative systems of point-like particles moving in vacuum (*non-relativistic exterior dynamical systems*), thus without any resistive force or contact interaction.

Since point-like particles are an approximation of the physical reality, the verifications of the EPR argument reported in Section 7 mandate the completion of Galileo symmetry and relativity for *non-relativistic interior dynamical systems* comprising extended particles in deep EPR entanglement, with ensuing resistive, as well as non-linear, non-local and non-Hamiltonian interactions (Figure 28).

The above need suggested the construction of the Lie-isotopic completion of the Galileo symmetry $\hat{G}(3.1)$ and relativity, called *Iso-Galilean isorelativity*, for the axiom-preserving representation of extended masses moving within physical media, thus experiencing resistive as well as contact interactions. Note that the axiom- preserving condition restricts the system to have a conserved total energy, as it is the case for the non-relativistic description of an isolated nucleus with contact internal forces.

Since conservative systems are an evident particular case of nonconservative/irreversible systems, the studies here considered required the construction of the broader Lie-admissible completion of the Galileo symmetry $\hat{G}^{>}(3.1)$ and relativity for the description of extended particles in nonconservative conditions, as it is the case for the non-relativistic representation of nuclear fusions.

Since non-relativistic studies are an evident pre-requisite for relativistic counterparts, we regret not to be able to review them to avoid a prohibitive length of this Overview. Nevertheless, we indicate for interested colleagues that non-relativistic studies were initiated in paper [19] of 1978 with the Lie-admissible completion of Galileo symmetry and relativity for non-conservative and Galileo- non-invariant systems.

The first direct study of Galileo's isosymmetry was done in Section 5.3, pages 225 on, of the 1983 monograph [22] formulated over conventional fields. These isotopies were then systematically studied and upgraded in the 1991 volumes [71] [72]. The formulation of Galilean isosymmetries with the full use of isomathematics was done in the 1995 monographs [23] [24] [73].

The above studies attracted the attention of Abdus Salam, founder and President of the *International Center for Theoretical Physics* (ICTP), Trieste, Italy, who invited Santilli in 1991 to deliver at his Center a series of lectures in the isotopies of the Galileo symmetry and relativity, said invitation being apparently the last one by Salam prior to his death. During his visit at the ICTP, Santilli wrote papers [147]-[153]. The notes from Santilli's lectures at the ICTP were collected by some of the attendees and published in volume [154] of 1992.

8.4.3. Special Relativity (SR). Mainly for the sake of notation, we recall that special relativity (SR) is defined on a Minkowski space $M(x, \eta, I)$ over the field of real numbers $\mathcal{R}$ with local coordinates $x = (x,^{\rho})$, $\rho = 1, 2, 3, 4$, $x^4 = ct$, metric $\eta = Diag.(1, 1, 1, -1)$, unit $I = Diag.(1, 1, 1, 1)$, and spacetime interval

$$(x - y.)^2 = (x - y)^{\mu}\eta_{\mu\ni}(x - y)^{\nu} =$$

$$(x_1 - y_1)^2 + (x_2 - y_2)^2 + (x_3 - y_3)^2 - (t_1 - t_2)^2c^2, \tag{116}$$

which is left invariant by the Lorentz symmetry $SO(3.1)$, the Lorentz-Poincaré symmetry $P(3.1)$ and its spinorial covering $\mathcal{P}(3.1)$.

The above methods uniquely and unambiguously characterize the following basic axioms for a *time-reversal invariant relativistic system of point-like particles and electromágnetic waves propagating in vacuum:*

AXIOM I: The speed of light c is the maximal causal speed for point-like particles propagating in vacuum

$$V_{max} = c \tag{117}$$

AXIOM II: The addition of speeds follows the relativistic law

$$V_{tot} = \frac{v_1 + v_2}{1 + \frac{v_1 v_2}{c^2}}. \tag{118}$$

AXIOM III: The dilation of time, the contraction of lengths, the variation of mass and the mass-energy equivalence follow the relativistic laws:

$$t' = \gamma\, t, \tag{119}$$

$$\ell' = \gamma^{-1}\, \ell, \tag{120}$$

$$m' = \gamma\, m, \tag{121}$$

$$E = mc^2 \tag{122}$$

where

$$\beta = \frac{v}{c}, \ \ \gamma = \frac{1}{\sqrt{1-\beta^2}}. \tag{123}$$

AXIOM IV: The frequency shift due to relative speed follows the law (for null aberration)

$$\omega_{rec} = \frac{\omega_{sou}}{\gamma\,[1-\beta\, cos(\alpha)]}. \tag{124}$$

The above axioms are hereon assumed to be exactly valid for the assumed time-reversal invariant systems.of point-like particles and electromagnetic waves in vacuum.

8.4.4. Special isorelativity (SIR). As a consequence of the widely assumed reduction of the entire universe to point-like particles, it is generally believed that special relativity and the constancy of the speed of light c, are valid for whatever conditions exist in the universe.

In the preceding sections we have shown that such a conception can be considered as being *approximately valid,* although it implies a number of insufficiencies, such as the inability to achieve exact representations of nuclear and molecular data due to the extended character of particles at a mutual distances of the order of $10^{-13}\ cm$ (relativistic interior dynamical problems).

When the reduction of the universe to point–like constituents is assumed as being *exactly valid,* it leads to serious inconsistencies that remain generally ignored by the mainstream physics community, although they need to be brought to the attention of the scholars in the field. For instance, the belief that electromagnetic waves propagating in water can be reduced to photons propagating in vacuum and scattering among the water molecules is afflicted by the following [24]:

Figure 29: *The light beam passing through a glass of water depicted in this figure is generally reduced to photons for the intent of maintaining special relativity in interior conditions. However, such a reduction is afflicted by the widely propagated (yet ignored) Inconsistencies 1) to 7) of Section 8.4.1 whose resolution mandates the completion of SR for interior dynamical conditions.*

INCONSISTENCY I: The reduction of light to photons cannot represent the angle of refraction of light in water, evidently because photons will scatter in all directions at the impact with the water surface.

INCONSISTENCY II: The reduction of light to photons cannot represent the decrease of the speed of light in water by $100,000\ km/s$, because calculations have shown that the scattering of photons among water molecules can at best represent a reduction of the speed of light of about $7,000\ km/s$.

INCONSISTENCY III: The reduction of light to photons cannot represent the propagation of light in water as a beam, again, because a beam of photons will scatter among the water molecules and disperse in water.

INCONSISTENCY IV: The reduction light to photons has no physical sense for infrared light or radio waves with $1\ m$ wavelength that experience the same phenomenology as that for electromagnetic waves at large.

INCONSISTENCY V: The reduction of light to photons scattering among water molecules, thus propagating in vacuum at speed c, cannot represent the Cherenkov light because said light can only occur when electrons travel *faster* than the local speed of light.

INCONSISTENCY VI: The reduction of light to photons cannot resolve the inapplicability of the relativistic sum of speeds in water, Eq. (118), since the sum of two light speeds in water does not yield the light speed in water.

INCONSISTENCY VII: The reduction of light to photons cannot be tested experimentally due to the lack of an inertial frame in water.

Following the construction of the axiom-preserving isotopies of the various branches of Lie's theory in the 1983 monograph [22], Santilli constructed the **axiom-preserving isotopies of special relativity for extended particles with interval** (25) in Ref. [60] for the classical part and Ref. [61] for the operator counterpart, with the first known construction of the the Lorentz-Isotopic symmetry $\hat{SO}(3.1)$, today known as the *Lorentz-Santilli isosymmetry,* with isotransformations (30).

Subsequently, Santilli conducted systematic studies [62]-[74] of the isotopies of all conventional spacetime symmetries resulting in the Lorentz-Poincaré- Santilli isosymmetry $\hat{P}(3.1)$ [67] and related isospinorial covering $\hat{\mathcal{P}}(.)$ [68][69] which are formulated on an iso-Minkowskian isospace first introduced in Ref. [60] (see Ref. [70] for detailed studies) $\hat{M}(\hat{x}, \hat{\Gamma}, \hat{I})$ over the isofield of isoreal isonumbers $\hat{\mathcal{R}}$ with isospacetime isocoordinates $\hat{x} = x\hat{I}$, $\hat{y} = y\hat{I}$, isometric correctly written as having elements in the isofield $\hat{\mathcal{R}}$

$$\hat{\Gamma} = \{\hat{\eta}_{\mu\nu}\}\hat{I} = (\hat{T}\eta)\hat{I}, \tag{125}$$

and isounit ($\hat{I} = 1/\hat{T} > 0$ given in Eq. (34), with invariant (25),which we now rewrite in the form

$$(\hat{x}-\hat{y})^{\hat{2}}=(\hat{x}-\hat{y})^{\mu}\hat{\times}\hat{\Gamma}_{\mu\nu}\hat{\times}(\hat{x}-\hat{y})^{\nu}=$$

$$=[(x-y)\hat{I}]\hat{T}(\hat{\eta}\hat{I})\hat{T}[(x-y)\hat{I}]=[(x-y)^{\mu}\hat{\eta}_{\mu\nu}(x-y)^{\nu}]\hat{I}=$$

$$[\frac{(x_1-y_1)^2}{n_1^2}+\frac{(x_2-y_2)^2}{n_2^2}+\frac{(x_3-y_3)^2}{n_3^2}-(t_1-t_2)^2\frac{c^2}{n_4^2}]\hat{I}, \qquad (126)$$

where: the multiplication of the interval by the isounit $\hat{I}$ is necessary for its value to be an element of the isofield; and the characteristic quantities $n_\rho,\ \rho=1,2,3,4$ are solely restricted by the condition of being positive-definite $n_\rho>0$, but possess otherwise an unrestricted functional dependence on all needed local variables (which shall be hereon tacitly assumed) such as local coordinates x, velocity v, momentum p, energy E, density ρ, temperature τ, pressure γ, frequency ω, wavefunctions ψ, their derivatives $\partial_r\psi$, etc. $n_\rho=n_\rho(t,r,v,p,E,\rho,\tau,\gamma,\omega,\psi,\partial\psi,.....)$.

A feature important for the understanding of this section is that all (non-singular) infinitely possible realizations of the iso-Minkowski isospace $\hat{M}$ on an isoreal isofield $\hat{\mathcal{R}}$ are locally isomorphic to the conventional space M over the reals $\mathcal{R}$. This property was first proved in the 1983 paper [60] (see Refs. [23] [72] for a detailed treatment) and can be seen from the fact that the Minkowski metric η is completed into the isometric $\hat{T}\eta$ while, jointly, the basic unit I is completed by the *inverse* amount $\hat{I}=1/\hat{T}$, thus preserving the original metric η.

Alternatively, one can see from Eq. (125) that the numeric value of the *Minkowskian* metric η is preserved under isotopies since $(\hat{T}\eta)\hat{I}\equiv\eta$.

Note the duality of the formulation, namely, the iso-Minkowski isospace can be first written on isospace over isofields (see the first line of interval (126)), in which case SR applies identically, and then *projected* on the conventional Minkowski space over a conventional field (see the third line of interval (126))where novelties appear.

Recall that the Minkowskian geometry represents a *homogeneous and isotropic 3-dimensional space* while, by comparison, the iso-Minkowskian isogeometry represents an *inhomogeneous space,* due to the local variation of the density, as well as an *anisotropic space,* due to the change of geometry with the change of the direction.

Consequently, axioms (117)-(124) of SR do not need the identification of the selected direction, while such an identification is necessary for the SIR due to the indicated variation of physical characteristics with the variation of the space direction.

Under the above clarifications, special isorelativity can be defined as the *axiom-preserving formulation of special relativity on iso- Minkowski isospaces over isoreal isofields.* Its universal LPS isosymmetry characterizes uniquely and unambiguously the following isoaxioms first formulated systematically in Refs, [72] of 1991 on conventional fields and completed in Ref.[24] of 1995 over isofields, here expressed for the selected k-direction, e.g., that of the third space component,

ISOAXIOM I: The speed of light within (transparent) physical media is given by the locally varying speed:

$$C = \frac{c}{n_4}. \tag{127}$$

ISOAXIOM II: The maximal causal speed within physical media is given by:

$$V_{max,K} = c\frac{n_k}{n_4}. \tag{128}$$

ISOAXIOM III: The addition of speeds within physical media follows the isotopic law:

$$V_{tot} = \frac{\frac{v1.k}{n_k} + \frac{v_{2.k}}{nn_k}}{1 + \frac{v_1 v_2}{c^2}\frac{n_4^2}{n_k^2}}. \tag{129}$$

ISOAXIOM IV: The isodilation of time, the isocontraction of lengths, the isovariation of mass and the mass-energy isoequivalence (isorenormalization) within physical media follow the isotopic laws:

$$t'_k = \hat{\gamma}_k\, t, \tag{130}$$

$$\ell'_k = \hat{\gamma}_k^{-1}\, \ell, \tag{131}$$

$$m'_k = \hat{\gamma}_k\, m, \tag{132}$$

$$\hat{E}_k = m\, V^2_{max,k} = \hat{m}_k c^2, \;\; \hat{m}_k = m \frac{n_k^2}{n_4^2}, \tag{133}$$

where, from Eqs. *(31)*

$$\hat{\beta}_k = \frac{v_3/n_k}{c_o/n_4}, \;\; \hat{\gamma}_k = \frac{1}{\sqrt{1-\hat{\beta}_k^2}}. \tag{134}$$

ISOAXIOM V: The frequency shift within physical media follows the isotopic law (for null aberration)

$$\omega_{rec} = \frac{\omega_{sou}}{\hat{\gamma}\,[1 - \hat{\beta}\; isocos(\hat{\alpha})]}. \tag{135}$$

Note that *the maximal causal speed in SIR is no longer given by the speed of light,* and it is given instead by value (128), because physical media are generally opaque to light, thus requiring the broader geometric notion derivable from the expression in $(k-4)$-space

$$\frac{dx_k^2}{n_k^2} - dt^2 \frac{c^2}{n_4^2} = 0. \tag{136}$$

Among a variety of verifications of Isoaxioms I to V, we indicate here the following representative examples:

8.4.4-I. Verification of SIR within liquid media. It is an instructive exercise for the interested colleague to verify that Isoaxioms I to V resolve all Inconsistencies I to7 (Figure 29) in the use of conventional axioms.

This includes the verification via isoaxiom (118) that the sum of two local speeds of light $C = c/n_4$ yields the local speed of light $C = c/n_4$. Moreover, water can be assumed to be homogeneous and isotropic for which $n_\mu = 1, \ \mu - 1, 2, 3, 4$ for which interval (126) becomes

$$(\hat{x} - \hat{y})^{\hat{2}} = (x - y)^2 \hat{I} =$$

$$= [(x_1 - y_1)^2 + (x_2 - y_2)^2 + (x_3 - y_3)^2 - (t_1 - t_2)^2 c^2] \hat{I}, \qquad (137)$$

in which case the maximal causal speed *in water* is the speed of light *c in vacuum,* by therefore providing the first known quantitative representation that *electrons can indeed travel in water faster than the local speed of light.*

Additionally, the iso-Minkowskian geometry [70] provides a geometric representation of the difference between the actual size d_{act} of an object in water and that perceived by an external observer d_{ext}. For this purpose, note that in water we have the value $\hat{I} = n_4^2 = 9/4 > 1$. The use of the basic law for isotopy [23] [24] $d_{act}^2 \hat{I} = d_{ext}^2 I$,, then yields the value

$$d_{ext} = n_4 d_{act} = \frac{3}{2} d_{act} \qquad (138)$$

which is essentially verified by visual inspection as one can see from Figure 30.

Intriguing similar properties occur for other external and internal characteristics. For instance we have a similar distinction between the observer time called *external time* t and related unit $I_t = 1$ (which is the time of the external observer), and the *intrinsic time* $\hat{t}$ with related isounit $\hat{I}_t$ (which is the time for an internal observer). Said two times are interconnected by the isotopic law $tI_t = \hat{t}\hat{I}_t$.

Figure 30: *A view of the apparent increase of size of objects submerged in water as seen by an outside observer. This increase is quantitatively represented by Special IsoRelativity (SIR) via Eq. (138) of Section 8.4.4-I with intriguing geometric implications.*

The above notion of *isotime* appears to be significant for biological structures [165] because we generally assume that the time felt by a seashell $\hat{i}$ is identical to our time t, while in reality the external and internal times may be different.

8.4.4-II. Verification of SIR within gaseous media. Recall that the Doppler law in vacuum, Eq. (124), can be written in first order approximation

$$z = \frac{\omega'}{\omega} - 1 \approx \pm\frac{v}{c}, \tag{139}$$

where the minus (plus) sign occurs for the source moving away (toward) the observer. The corresponding expansion for the iso-Doppler law (135) yields the expression

$$\hat{z} = \frac{\omega'}{\omega} - 1 \approx \pm\frac{v_k}{c}\frac{n_4}{n_k}, \tag{140}$$

where the factor n_4/n_k can be expanded in terms of $(c/v_k)d$, d being the

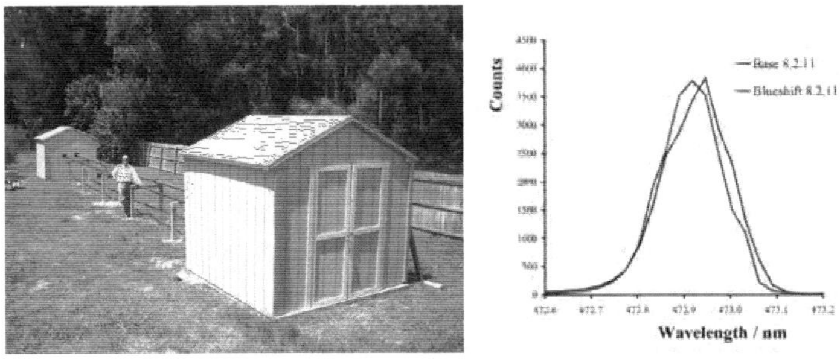

Figure 31: *A view on the left of the* $60'$*-long Isoshift Testing Station that established the energy loss (gain), thus frequency decrease (increase) without any relative motion of a blue laser light passing through air at* 1000 *psi and at* $-10^0\ C$ *(*$+150^0\ C$*) called isoredshift (isoblueshift), which was first predicted in the 1991 Refs. [71] [72], and first detected in the 2010 paper [155] via the scan shown in the right view.*

travel of light in the medium

$$\frac{n_4}{n_k} \approx 1 \pm S(\rho, \tau,) \frac{c}{v_k} d, \tag{141}$$

where S s a function of the local density ρ, temperature τ and possibly other local variables.

The iso-Doppler law (135) then assumes the form in first order approximation, first derived in the 1991 Refs. [71] [72],

$$\hat{z} = \frac{\omega'}{\omega} - 1 \approx \pm \frac{v}{c} \pm S(\rho, \tau, ...) d, \tag{142}$$

which represents two independent frequency shifts, the conventional Doppler's shift $\pm v/c$ due to relative motion between observer and source, and the isoshift $\pm S(\rho, \tau, ...)d$ due to loss (absorption) of energy $E = h\nu$ to (from) a gas. at temperatures lower (bigger) than $0\ C$, called *isoredshift (isoblueshift)*, in which the prefix "iso" denotes the sole possible derivation via isomathematics.

Following a decade of failed attempts to locate a physics laboratory interested in proving or disproving the prediction of the isoshift [72], Santilli conducted systematic tests in collaboration with the technicians of *Magnegas Corporation* at its former laboratories located at 150 Rainville Rd, Tarpon Springs, Florida. An *isoshift Test Station* (left view of Figure 31) was built consisting of a front (rear) air-conditioned cabins containing a blue laser light (wavelength analyzers), the two cabins being interconnected by a $60\ ft$ long tube containing air at $1,000\ psi$. The air temperature was varied from $-30\ C$ to $+200\ C$ via commercially available cooling or heating means..

Systematic tests reported in the 2010 paper [155], established that a blue laser light loses (gains) energy, thus decreasing (increasing) its frequency, when passing through air contained in the indicated tube at $1,000$ psi and $-10\ C$ ($+150\ C$).

Since none of the measured frequency shifts occurred with any relative motion between the source and the observer, tests [155] established the existence of the isoshift as predicted in the 1991 monograph [72].

Systematic additional measurements were conducted in the USA and Europe via the best available wavelength analyzers by following Sunlight with a telescope from the Zenith to the horizon. These tests established that the redness of the Sun at the horizon (left view of Figure 32) is due to a loss of energy to our atmosphere such to cause the isoredshift of the blue light into the red light (right view of Figure 32) [157] (see also lecture f). It should be noted that the scattering of photons among the molecules of our atmosphere is basically insufficient to represent the large redshift from the blue all the way to the red.

8.4.4-III. Verification of SIR in astrophysics. Measurements [159] provided an experimental verification on Earth of Zwicky's

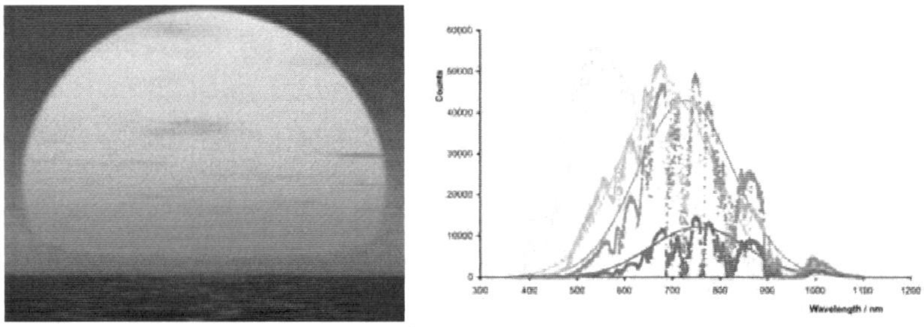

Figure 32: *In order to maintain the validity of special relativity within our inhomogeneous and anisotropic atmosphere, it is generally believed that the redness of Sunlight at Sunset is due to the absorption of blue light resulting in the residual red light, against evidence known since Newton's time that blue light is the most penetrant and red light is quickly absorbed by media. Systematic measurements [159] done in the U.S.A. and in Europe established that the redness of the Sun at Sunset (left view) is due to loss of energy to our atmosphere with ensuing isoredshift of blue light into the red red light (right view).*

hypothesis of *Tired Light* [160], according to which galactic light loses energy to the very cold intergalactic medium (mostly composed by Hydrogen at absolute zero degree temperature) in a way essentially proportional to the covered distance d in a static universe without motion of galaxies at such a velocity v/c to provide a contribution to the redshift.

In particular, SIR provides a direct representation of Hubble's law with the identification of the S quantity of Eq. (142) with the Hubble constant H_0 without any contribution from the Doppler shift

$$\hat{z} = \frac{\omega'}{\omega} - 1 = -\frac{H_0}{c}d. \tag{143}$$

The above isolaw provides an excellent representation of astrophysical data on cosmological redshift as shown in Ref.[161], including a proportionality from the distance, the representation of the very large redshift of light from galaxies at the end of the known universe without any need for superluminal speeds of entire galaxies or the hyperbolic and unverifiable assumption that space itself is expanding.

The numerical representation of internal galactic redshift anomalies occurs via the use of the isoredshift for star light propagating through the cold peripheral intergalactic medium, and the isoblueshift for star light passing trough hot intergalactic medium near a black hole [162].

The acceleration of the cosmological redshift with the distance d can be beautifully represented via a gravitational contribution to the redshift of galactic light passing near stars or galaxies in their long travel to Earth [163].

The above results set the foundations of the new *Tired Light Cosmology representing a static universe in total EPR entanglement* (Figure 1) according to the view preferred by Einstein, Hubble, Hoyle, Zwicky, Fermi, and others who died without accepting the expansion of the universe.

Note that for about one century the cosmological redshift z has been interpreted to be a direct "measurement" of the speed v of galaxies in an expanding universe according to the SR law (139), while in reality the identification $z = v/c$ is an *assumption* due to the existence of the Tired Light and other interpretations of the cosmological redshift.

SIR has established that the identification $z = v/c$ is an experimentally unverifiable *assumption* since the cosmological redshift z can be equally interpreted via Zwicky's Tired Light without any expansion of the universe, and via other models such as that of the *Tired Time* [85].

Note that the assumption $z = v/c$ implies

$$\frac{v}{c} = \frac{H_o}{c} d, \tag{144}$$

from which

$$v = H_o d, \tag{145}$$

which establishes the *radial* character of the conjectured expansion of the universe necessarily implying Earth at its center [163].

Following one century of oblivion of Einstein's rejection of the expansion of the universe as well as the oblivion of Einstein's view that "quantum mechanics is not a complete theory, " it is hoped that cosmologists will compare the new Tired Light Cosmology with EPR entangled universe [155]-[162] with the forgotten conceptual, geometrical and physical insufficiencies or sheer inconsistencies of the unverifiable conjecture of the expansion of the universe [163] [164] (see also debate [76] for details).

8.4.4-IV. Verification of SIR with the mean-life of unstable hadrons. Ref.[166] of 1964 suggested that non-linear and non-local effects in the interior of hadrons caused by their high density can manifest themselves in the outside via deviations of the behavior with energy of their mean lives τ from time evolution law (119), i.e.,

$$\tau' = \frac{\tau}{\sqrt{1 - \frac{v^2}{c^2}}}. \tag{146}$$

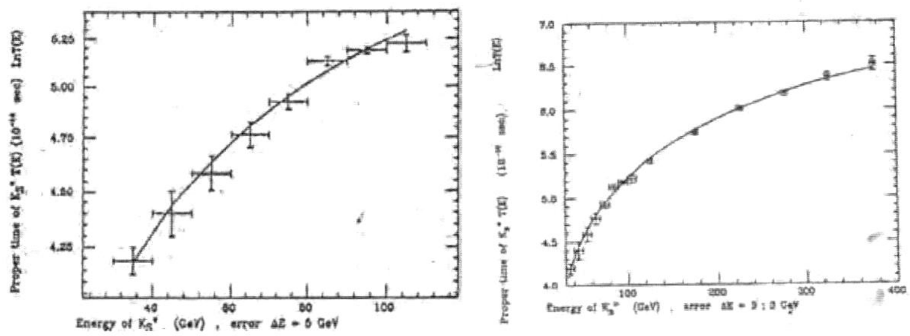

Figure 33: *In this figure, we show the exact fit by special isorelativity experimental data [168] showing deviations from special relativity law (119) in the behavior of the mean-life of Kaons from* 1 *to* 100 *GeV (left fit) as well as experimental data [168] joined with those of Ref. [169] claiming the confirmation of special relativity law (119) between* 100 *to* 400 *GeV (right fit) [170]-[172].*

Numerous generalizations of said law were then proposed, but Ref. [167] showed that, in view of the universality of the Lorentz-Santilli isosymmetry $\hat{SO}(3.1)$ for all possible symmetric space-tines (116), all said generalized time evolution laws are particular cases of the SIR isolaw (130), i.e.

$$\tau' = \frac{\tau}{\sqrt{1 - \frac{v_k^2}{c^2}\frac{n_4^2}{n_k^2}}}. \tag{147}$$

since they can be all obtained via different expansions of the term n_4^2/n_k^2 in different variables and with different truncations. Consequently, Ref. [167] provided a significant clarification for experiments since they can be all restricted to test isolaw (130).

Experiments on the indicated prediction of deviation from law (146) were conducted in 1983 [168] for the behavior of the unstable kaons with speed and established deviations from law (119), consequently in favor of isolaw (130) from 1 to 100 GeV. A counter-experiment was done in 1987 [169] claiming the confirmation of law (119) between the *different* range of 100 to 400 GeV.

Refs. [170] [171] showed that the data from both experiments [168] and [169] can be exactly fit with the iso-Minkowskian geometry of relativistic hadronic mechanics [60] [70] (Figure 33). The 1992 Ref. [172] confirmed the findings of Refs. [170] [171], by indicating flaws in the theoretical elaboration of the form factors of counter-experiments [169]. A detailed presentation is available in Ref. [74], Vol. IV, Section 9.

It is unfortunate for our scientific knowledge that. according to the official position of the editors of the journals of major physical societies, SR is claimed to be exactly valid in the interior of hadrons following the 1987 counter-experiment [169] despite the fact that the deviations from SR laws remained established between 100 to 400 GeV and in complete oblivion of refs. [170] [171], while requests made to various particle physics laboratories over decades to run the experiment again rom 1 to 400 GeV have been generally discredited.

8.4.4-V. Verification of SIR in the Bose-Einstein correlation. As it is well known (see, e.g., Ref. [173]), the fit of the experimental data of the *two*-point correlation function of the Bose-Einstein correlation for proton-antiproton annihilation requires *four* different arbitrary parameters of unknown origin or meaning called the *chaoticity parameters*. It has been shown in Refs. [85] [86] that this occurrence is clear evidence on the lack of exact character of SR for proton-antiproton annihilation for a number of reasons, including the fact that the vacuum expectation value of the two-point correlation operator $C_{2\times 2}$ is given by the known quantum mechanical expression

$$< C_{2\times 2} >=< \psi|C_{2\times 2}|\psi >, \tag{148}$$

which, being two-dimensional and real-valued, can at best allow *two* parameter following due manipulation of the basic axioms.

By contrast, the use of relativistic hadronic mechanics [24] implies the

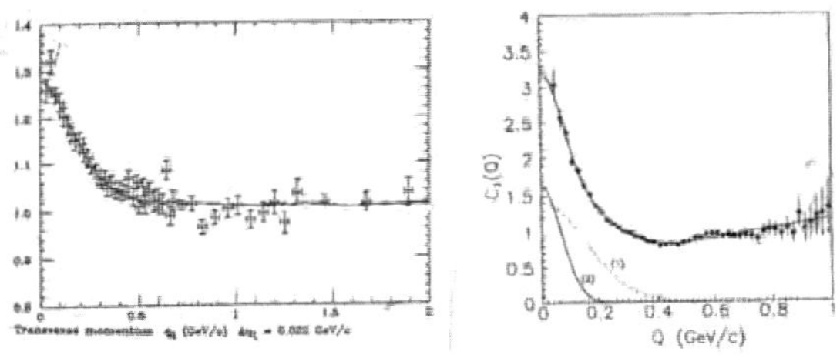

Figure 34: *The fit of experimental data for the two-point correlation function of proton-antiproton annihilation in the Bose-Einstein correlation requires 'four' arbitrary parameters of unknown origin [173]. Refs. [85] [86] (see also review [174]) have shown the inapplicability of special relativity for the proton-antiproton annihilation in favor of the exact fit via special isorelativity at high energy (left plot) and low energy (right plot) from which the four characteristic quantities n_μ of the iso-Minkowskian geometry represent the very elongated fireball of proton-antiproton annihilations, Eqs. (150)-(151).*

following isoexpectation value

$$\hat{<}C_{2\times 2}\hat{>} = <\hat{\psi}|\hat{\times} C_{2\times 2}\hat{\times}|\hat{\psi}> = <\hat{\psi}|\hat{T}_{2\times 2} C_{2\times 2}\hat{T}_{2\times 2}|\hat{\psi}> . \tag{149}$$

It is then easy to see that the positive-definite 2×2-dimensional isotopic element $\hat{T}_{2\times 2} > 0$ does indeed allow in its non-diagonal realization the introduction of four characteristic quantities n_μ^2, $\mu = 1, 2, 3, 4$ from first axiomatic principles without any adulteration, whose value is fit from the experimental data resulting in the values (Ref. [85], Eqs. (10.27a), page 127, Ref. [86], Table I, page 441, and review [174])

$$n_1^2 = n_2^2 = 10.666, \quad n_3^2 = 0.355,$$

$$n_1 = n_2 = 3.265, \quad n_3 = 0.595 \tag{150}$$

$$n_4^2 = 0.429, \quad n_4 = 0.654. \tag{151}$$

One can see from values (150) that the fit of the experimental data (Figure 34) characterizes the known prolongated toroid of the proton-antiproton fireball, while value (151) characterizes its density (for details, see Ref. [75]. Volume IV, Section 10, page 802-809 and review [174]).

8.4.4-VI. Verification of SIR with superluminal speeds.

8.4.4.VIA. Solution of the historical Lorentz problem.

The most important implication of the axiom-preserving isotopies of the Lorentz symmetry achieved in the 1983 paper [60] is the *solution of the historical Lorentz problem,* namely, the invariance of the locally varying speeds of light $C = c/n$ that Lorentz could not solve due to the linearity cf Lie's theory, by being therefore forced to restrict his invariance to the particular case of a constant c.

In the author's view, the most important implication of the EPR argument [1] applied to isosymmetry [60] can be expressed with the following property from Ref. [175] [176]: :

LEMMA 8.4.1. The axioms of Einstein's special relativity provide a representation invariant over time of speeds of light through transparent media that can be arbitrarily bigger or smaller than the speed of light in vacuum,

$$C = \frac{c}{n_4} \lesseqqgtr c. \tag{152}$$

PROOF:
Isosymmetries $\hat{SO}(3.1)$, $\hat{P}(3.1)$ and $\hat{\mathcal{P}}$ of isospacetime(126) [60]-[70] (see Refs. [24] and [213] for a review) require no restrictions on the value and functional dependence of the characteristic quantities of the medium, except for the condition of being positive definite, $n_\mu > 0$, $\mu = 1, 2, 3, 4$. Consequently, said isosymmetries provide a characterization of arbitrary local speeds of light, Eq. (152), in a way fully compatible with the axioms of SR because said isosymmetries are locally isomorphic to the corresponding conventional spacetime symmetries $SO(3.1)$, $P(3.1)$, $\mathcal{P}$. The

proof of the invariance over time of arbitrary speeds (152) can be done via a simple isotopy of the invariance over time of the speed of light c under the conventional symmetries $SO(3.1),\ P(3.1),\ \mathcal{P}$.
Q.E.D.

Alternatively, SR can be fully realized on the Minkowski-Santilli isospace [70] $\hat{M}(\hat{x}, \hat{\Gamma}, \hat{I})$ over the isoreal isofields $\hat{\mathcal{R}}$ with isounit $\hat{I} = 1/\hat{T} > 0$, in which case the line element given by the top line of Eq. (126).

The realization of the isocoordinates

$$\hat{x}^{\mu} = \frac{x^{\mu}}{n_{\mu}}\hat{I}, \tag{153}$$

then yields isosymmetries $\hat{SO}(3.1),\ \hat{P}(3.1),\ \hat{\mathcal{P}}$ with arbitrary speeds (152).

The propagation of light within transparent liquids with densities $n_4 > 1$ and light speeds *smaller* than that in vacuum, $C = c/n_4 < c$, has been known for centuries (Section 8.4.4-I), e.g., for the case of water with density characterized by $n_4 = 1.5$ for which we have (Figures 29, 30)

$$C = \frac{c}{1.5} = 0.666c = 200K\ km/s. \tag{154}$$

Note that the characteristic quantity n_4 provides a *geometric characterization* of the density of the medium whose actual value is given by $D = 1/n_4$.

Note also Inconsistencies 1) to 7) in the event mainstream physicists attempt to maintain in water the speed of light in vacuum via the usual reduction of electromagnetic waves to photons.

Recall that the Cherenkov light we see in the pools of nuclear power plants establishes the propagation of electrons in water at speed *bigger* than the local speed of light.

SIR has generalized the above Cherenkov effect to hyperdense physical media with geometrized densities $n_4 < 1$ resulting in arbitrary speeds

$C = c/n_4 > 1$. This result is implicit in all experimental fits of particle physics experiments to date via the SIR (e.g., Figures 33, 34). The result is also implicit in all structure models of hadrons, nuclei and stars based on the EPR completion of quantum into hadronic mechanics [36] [74].

As an illustration, recall from Section 8.2.2 that the *sole* known possibility to represent *all* characteristics of the neutron in its synthesis from the hydrogen is that the energy of the isoelectron is isorenormalized according to Eq. (36) from $m_e = 0.511\ MeV$ to $m_{\hat{e}} = 0.511 + 0.784 = 1.295\ MeV$, by therefore yielding the expected geometrized density of the proton

$$n_4 = \frac{0.511}{1.295} = 0.394 < 1, \tag{155}$$

whose order of magnitude is confirmed by the geometrized density of the proton-antiproton fireball in the Bose-Einstein correlation (Ref. [85], Eqs. (10.27a), page 127).

This implies that *the isoelectron is rotating inside the hyperdense proton (Figure 11) with the superluminal tangential speed*

$$C = \frac{c}{0.394} = 2.538\ c. \tag{156}$$

Superliuminal speeds are generally obtained within the π^0 meson [20] and hold for all remaining hadrons since they all have masses bigger than that of the π^0, while having essentially the same size of the π^0. Consequently, SIR suggests that *particles travel at a superluminal speed that begins with the interior of the π^0, and increases with the increase of the mass of the hadrons in a way parallel to the progressive regaining of Einstein's determinism with the increase of the density of hadrons* [210]-[214].

8.4.4.VIB. Geometric locomotion. It may be of some interest to mention the *mathematical* prediction by SIR of spaceships traveling at speeds much bigger than the speed of light in vacuum which speeds are evidently necessary for any interstellar travel. The prediction was submitted in Section 4.3.3, page 281 of the 2006 monograph [29] under the name of *geometric locomotion,* also known as *isolocomotion* due to the need of its treatment via isomathematics. Isolocomotion is an intrinsic feature of the iso-Minkowskian geometry [70] and its Lorentz-Santilli isosymmetry [60].

Recall that the main notion of any geometry dating back to Euclid is the distance d between two points, which we write $d1$, where 1 is the unit of measurement (as well as the unit of the basic numeric field), e.g. $1\ m$. Recall the preservation of numeric values under axiom-preserving isotopies [24], which we write $d\ 1 \equiv \hat{d}\hat{1}$. Isolocomotion is based on means capable of altering the local geometry in such a way to *increase* the value of the isounit $\hat{1}$ in the desired direction, with consequential *decrease* of the distance $\hat{d}$ in the selected direction. Isolocomotion then occurs without any possible geometric singularity, such as instantaneous accelerations, sharp changes of direction, or arbitrary speeds because the interior observer in the iso-Minkowski isospace is at rest since locomotion is achieved via the change of the geometry in its environment. By contrast, an observer in our external Minkowski space may see the distance d being covered at a multiple the speed of light c.

Alternatively, the aim of isolocomotion is to achieve arbitrary speeds without violating SR. This is *mathematically* achieved by turning a spaceship into rather complex *interior* conditions via the mutation of the surrounding geometry, with ensuing local applicability of SIR and related arbitrary speeds.

8.4.4.VIC. Interstellar travel. Recall that no interstellar travel is nowadays possible because of: 1) The need for superluminal speeds, 2) The impossibility of carrying along the necessary fuel, 3) The ability of changing directions at superluminal speeds to avoid collisions, 4) The need for virtually instantaneous communications, and other requirements.

The possible achievement of superluminal speeds has been discussed in the preceding section. Studies on a future resolution of the fuel problem have been initiated in Refs. [29] [175]. The main result is that quantitative studies for future interstellar travels can indeed be conducted provided that space is conceived as a universal substratum (ether) with an extremely big energy density for the characterization and propagation of particles and electromagnetic waves (Section 8.2.3).

In different words, the sole known possibility for interstellar travel is that the needed fuel has to be extracted from the ether.

The complexity of the problem soon emerges when considering the need for *two* co-existing ethers, one with positive energy for the characterization of matter, and one with negative energy for the characterization of antimatter [29]. This structure is of such a complexity that, in the author's view, can be best represented via the hyperstructural branch of hadronic mechanics [24].

Despite its complexity, the above conception of a universal substratum implies that immense values of positive and negative energy are available everywhere in the universe. Future attempts at the realization of isolocomotion are then reduced to engineering means for the directional transfer of *negative* energy from the ether to the spaceship. Under the indicated conditions, isolocomotion occurs via matter-antimatter repulsion.

The third problem (how to avoid collisions with astrophysical objects) is perhaps the most complex of all problems connected with interstellar travel since its sole solution is a change of the very structure of the spaceship in such a way to pass through planets without any damage, as shown in reports by the U. S. Navy of UFO entering in the sea at high speeds without causing any wave. Conceivably, the alteration of the geometry of the ether can *mathematically* be such to produce the needed "de-materialization."

In short, the ether appears to have a truly fundamental role for interstellar travel due to various reasons treated in Ref. [175], as a result of which the ether is considered by an increasing number of scientists as the most important frontier of the third millennium.

In closing, we point out that the entire content of this section, including superluminal speeds, is fully compatible with the axioms of Einstein's special relativity, only realized in their most general possible form [175] [176].

An additional study of the implications of the EPR argument for superliuminal communications was presented at the 2020 Teleconference [0] by Y. F. Chang from China and it is available in Ref. [218].

8.5. Application of EPR completions to exterior gravitational problems.

8.5.1. Universal invariance of Riemannian intervals.

As it is well known, Lie's theory can only provide the invariance of *linear* theories and this explains the reason for the inability to construct Lie symmetries of Riemannian intervals due to their non-linearity.

A primary motivation for the construction of the *non-linear* completion of Lie's theory into the covering Lie-Santilli isotheory [21]-[25] [50] [55] [201] [226] has been the construction of the universal isosymmetry of Riemannian intervals.

The above objective was studied for the particular case of isointerval (126) in which the isometric $\hat{\eta}_{\mu\nu}$ coincides with a Riemannian metric $g(x)$, and resulted in the following basic isosymmetries (see the review in Ref. [213]): 1) The iso-Lorentz symmetry, nowadays called the *Lorentz-Santilli (LS) isosymmetry* $\hat{SO}(3.1)$ first achieved at the classical level in the 1983 paper [60] with operator counterpart in Ref. [61];

2) The *iso-Poincaré isosymmetry*, nowadays called the *Lorentz-Poincaré-Santilli (LPS) isosymmetry* $\hat{P}(3.1)$, first achieved in the 1993 paper [67] written at Moscow State University; and

3) The spinorial covering of the LPS isosymmetry $\hat{\mathcal{P}}(3.1)$, first achieved

at the 1994 paper [68] written at the JINR, Dubna, Russia (see also the 1995 paper [69] published in China).

The understanding of the remaining parts of this section requires a technical knowledge that the isotopic symmetries formulated on iso-Minkowskian isospaces over isofields are locally isomorphic to the corresponding conventional symmetries formulated on a Minkowski space over a numeric field, i.e., $\hat{SO}(3.1) \approx SO(3.1), \quad \hat{P}(3.1) \approx P(3.1), \ \hat{\mathcal{P}}(3.1) \approx \mathcal{P}(3.1)$.

8.5.2. Isogeometric unification of special and general relativity. As indicated earlier, the *iso-Minkowskian geometry* [60] [70] includes as particular cases all possible geometries with symmetric space-time intervals (126), thus including the Minkowskian, Riemannian, Fynslerian and other geometries.

Consequently, the iso-Minkowski geometry is particularly suited for the *isogeometric unification of Einstein's special and general relativities under the universal LPS isosymmetry,* namely, a unification based on isomathematics while maintaining identically Einstein's field equations

$$R_{\mu\nu} - \frac{1}{2} R_{\mu\nu} R = 0. \tag{157}$$

Consider an arbitrary non-singular Riemannian metric $g(x)$ where x are the conventional space-time coordinates. Its *identical* reformulation via isomathematics requires the decomposition of $g(x)$ into the product of the Minkowski metric η multiplied by a 4×4 positive-definite *gravitational isotopic element* $\hat{T}_{gr}(x)$ [177] [178],

$$g(x) = T_{gr}(x) \times \eta = \hat{\eta}_{gr}. \tag{158}$$

The resulting iso-Minkowskian isospace $\hat{M}(\hat{x}_{gr}, \hat{\eta}_{gr}, \hat{I}_{gr})$ is then formulated over an isoreal isofield $\hat{\mathcal{R}}$ with *gravitational isounit* given by the *inverse* of the gravitational isotopic element,

$$\hat{I}_{gr}(x) = 1/\hat{T}_{gr}(x), \tag{159}$$

and local isocoordinates

$$\hat{x}_{gr} = x\hat{I}_{gr}. \tag{160}$$

Exterior general isorelativity (EGIR) [177]- [182] can be defined as *the formulation of general relativity on isospace $\hat{M}(\hat{x}_{gr}, \hat{\eta}_{gt}, \hat{I}_{gr})$ over an isoreal isofield $\hat{\mathfrak{R}}$ under the condition that its projection on a Riemannian space over a conventional field recovers general relativity uniquely and identically.* Consequently, by construction, isogravitation is a mere reformulation of general relativity via the use of isomathematics. The terms *exterior isogravitation* are used to recall that general relativity describes the exterior gravitational field in vacuum, in preparation of the complementary interior gravitational problem studied in the next section.

Note that, in view of the dependence of the isometric $\hat{\eta}(x)$ on spacetime coordinates, the iso-Minkowski isogeometry is formulated with the entire mathematical machinery of the Riemannian geometry, although expressed in terms of the isodifferential isocalculus [70].

Note also that isogravitation provides the reformulation of general relativity with the axioms of *special* relativity in their isotopic formulation, Isoaxioms I to V.

The isogeometric unification of special and general relativities is then assured by the fact that isogravitation is an identical reformulation of general relativity while admitting special relativity at the simple limit

$$Lim\hat{I}_{gr}\ I. \tag{161}$$

8.5.3. Resolution of century-old controversies in gravitation? As it is well known to experts in gravitation, although rarely admitted, general relativity has been plagued by a host of controversies that have not been resolved in about one century of studies, such as the apparent incompatibility of general relativity with special relativity, quantum mechanics, grand unifications and other 20th century theories.

In the author's view, protracted *physical* controversies are generally due to the fact that the used *mathematics* is insufficient for the solution of the problem considered. For the case of general relativity, it has been shown that the origin of the controversies can be reduced to the incompatibility of the Riemannian geometry with the axiomatic structure of 20th century theories, since the latter are all defined on flat spaces, while the former is defined on a curved space.

Isogravitation (GIR), that is, the reformulation of general relativity via the iso-Minkowskian geometry, appears to offer realistic possibilities of resolving the indicated century-old controversies, as illustrated by the following comments:

1) Isogravitation is isoflat, namely, it is flat on isospace over isofield. This important feature can be seen from the fact that the isogravitational isometric $\hat{\eta}_{gr}$ is given by the Minkowski metric η multiplied by the isotopic element $\hat{T}_{gr}$ according to Eq. (158), while jointly the basic unit of the Minkowski geometry $I = Diag.(1,1,1,1)$ is completed by the *inverse* amount, Eq. (150), resulting in no curvature. In the author's view, the isoflatness of GIR is important for the achievement of compatibility between gravitation and 20th century theories for the indicated reason that the latter are formulated in flat spaces.

2) Invariance under isosymmetries that are locally isomorphic to the corresponding conventional symmetries (Section 8.5.1). Recall that another reason for the incompatibility of GR with 20th century theories is the invariance of the former under the Lorentz and Poincaré symmetries compared to the lack of invariance for GR. It is then evident that the reformulation of gravitation in a form admitting symmetries locally isomorphic to the conventional symmetries provides serious support for the compatibility of isogravitation and 20th century theories.

3) Unique and unambiguous limit of isogravitation into special relativity [177]. Recall that the limit of general into special relativity has remained controversial for a century due to numerous reasons [181], such as the impossibility of recovering from the Riemannian geometry the *Poincaré symmetry* of special relativity, let alone its generators (conserved quantities). These controversies appear to be resolved by isogravitation due to its formulation via the axioms of *special* relativity. Additionally, SR is recovered from GIR in full via simple limit (161).

4) Unique and unambiguous operator formulation of gravitation [177]. An additional reason for controversies is the lack of a clear and unambiguous quantum mechanical formulation of GR. This is due to the fact that GR is a *non-canonical* theory whose consistent operator image is then given by *non-unitary* theories. But the latter theories formulated in conventional fields violate causality as indicated various times in this Overview, resulting in the lack of a unique, physically consistent operator image of GR. This additional, century-old controversy is resolved by the fact that the operator image of GIR is uniquely and unambiguously given by relativistic hadronic mechanics [36] when characterized by the gravitational isounit $\hat{I}_{gr}$, E q. (159). Additionally, relativistic hadronic mechanics is *isounitary,* that is, unitary on isospaces over isofields, thus recovering causality.

5) Unique and unambiguous grand unification [178]. Recall the impossibility to achieve a consistent grand unification of gravitation with other interactions beginning with Einstein's own failed attempts. In addition to a number of problematic aspects [181], this impossibility is primarily due to the *curvature* of the Riemannian geometry because, when combined with electromagnetic and/or weak interactions, curvature causes the collapse of their axiomatic structure, beginning with the loss of spacetime, gauge and other symmetries. This additional century-old controversy appears to be resolved by isogravitation because of its isoflatness, as studied in detail in monograph [29]. Note that the resulting grand unification required, for consistency, the addition of the gravity of antimatter via the isodual cf completion of charge conjugation (Section 7.7).

6) Historical objections against the curvature of space. As it is well known, when looking at the Sun at Sunset, the Sun is already below the horizon due to the *refraction* of Sun light in our atmosphere without any possible curvature of space. The historical, well known (but rarely mentioned) objection against the curvature of space is that the bending of light passing near our Sun is due to its refraction of light within the Sun chromospheres. Additional controversies occur from the fact that space is assumed to be empty. It is therefore counter-intuitive for a number of physicists to accept the idea that the curvature of an empty space can control the trajectory of large planets such as Jupiter. Additional controversies on curvature can be found in the debate [192].

7) Lack of time-invariant numerical predictions. Recall that a majestic feature of special relativity is the preservation over time of numerical prediction due to the *invariance* of the theory under the Lorentz-Poincaré symmetry. The canonical character of SR then assures the *uniqueness of the space-time metric for all experimental verifications.* As it is well known to historians (but also rarely spoken), the lack of Lie symmetries in the Riemannian geometry mandated the replacement in GR of the notion of invariance with that of *covariance.* However, such a replacement triggered a number of controversies, the first controversy being due to the the lack of uniqueness of the Riemannian metric for all experimental verifications in exterior conditions, with ensuing lack of final experimental results. Secondly, the use of covariance instead of invariance implies that numerical predictions of GR are not preserved over time, with the ensuing additional reason for the lack of final character of experimental verifications [161]. Additional controversies have occurred for experimental verifications of GR due to the apparent *ad hoc* selection of the PPN approximation for the selected experiment. It appears that GIR offers realistic possibility of resolving the indicated controversies on experimental verifications in case seeded in a receptive scientific environment (for additional controversies, see next section and Ref. [182]).

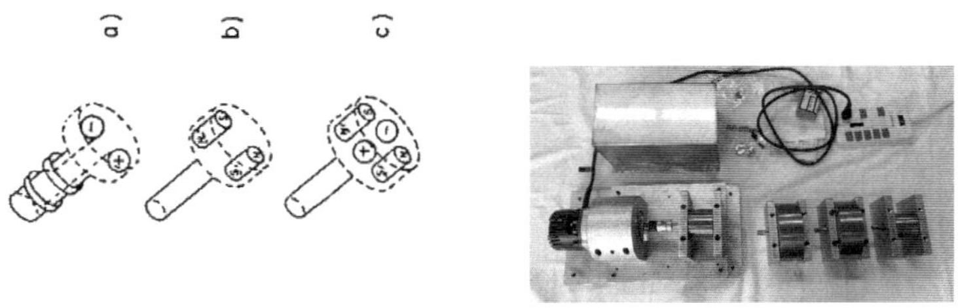

Figure 35: *On the left of this figure, we reproduce the proposal in page 145 of the 1974 MIT paper [185] to test the Poincaré hypothesis that the exterior gravitational field of a mass is entirely due to the electric and magnetic fields of the charged constituents in high dynamical conditions. In the right of this figure, we show the Gravity Generator Equipment (GGE) manufactured by Hadronic Technologies Corporation to test the Poincaré hypothesis in the expectation of its conduction by a physics laboratory with a sufficiently sensitive gravity meter.*

8.6. Application of EPR completions to interior gravitational problems.

8.6.1. Exterior and interior gravitational problems. In the first part of the 20th century, dynamical problems were called *exterior problems* when dealing with point-particles in vacuum and *interior problems* when dealing with particles in the interior of physical media. For instance, Schwartzschild wrote *two* important papers: the first paper [183] in the *exterior gravitational problem* that became justly famous, and the second paper [184] on the *interior gravitational problem* that continues to remain vastly ignored because not compatible with Einstein general relativity.

This view is essentially based on the fact that the reduction of matter to *point-like* constituents according to quantum mechanics, essentially eliminates any major distinction between exterior and interior problems.

The admission of the physical reality that particles in general, and hadrons in particular, are *extended,* and the ensuing verifications [210]-[214] of the EPR argument [1], reinstate the structural difference between exterior and interior gravitational problems adopted by Schwartzschild [183] [184[, due to the very complex interactions occurring for particles in interior conditions, ultimately reducible to deep EPR entanglements (Figure 1 and Section 7.2.3), which internal interactions are completely absent for the same particles in exterior conditions.

8.6.2. Interior General Isorelativity (IGIR). As noted in Section 8.5, general isorelativity (GIR) is characterized by an isometric solely dependent on space-time isocoordinates $\hat{\eta}(\hat{x})$, in which case the theory can solely represent the exterior gravitational field.

In general, the isometric has an unrestricted dependence on all needed local variables [70]

$$\hat{\eta} = \hat{\eta}(x, v, p, E, \mu, \tau, \rho, ...) = \hat{T}_{gr}(x, v, p, E, \mu, \tau, \rho, ...)\eta. \tag{162}$$

Consequently, isogravitation may indeed allow the study of interior gravitational problems, resulting in a theory called *interior general isorelativity* (IGIR).

One of the best illustrations of IGIR is given by the *iso-Dirac isoequation,* also known as *Dirac-Santilli isoequation,* on an iso-Minkowskian isospace $\hat{M}(\hat{x}, \hat{\eta}, \hat{I})$ first introduced in Ref. [70] of 1995

$$(-i\hat{I}\hat{\eta}^{\mu\nu}\hat{\gamma}_\mu\partial_\nu + mC)|\hat{\psi}(\hat{x}) >= 0. \tag{163}$$

In this case, the *Dirac-Santilli isomatrices* $\hat{\gamma}\hat{I}$ are given by

$$\hat{\gamma}_k = \frac{1}{n_k}\begin{pmatrix} 0 & \hat{\sigma}_k \\ -\hat{\sigma}_k & 0 \end{pmatrix}, \quad \hat{\gamma}_4 = \frac{i}{n_4}\begin{pmatrix} I_{2\times 2} & 0 \\ 0 & -I_{2\times 2} \end{pmatrix}, \tag{164}$$

where $\hat{\sigma}_k$ are the *regular Pauli-Santilli isomatrices* used for the EPR verification [211] (see realization(39) via Bohm's hidden variables and Section

3.3 of Ref. [213] for the general case), with *anti-isocommutation rules*

$$\{\hat{\gamma}_\mu,\hat{\gamma}_\nu\} = \hat{\gamma}_\mu \hat{T} \hat{\gamma}_\nu + \hat{\gamma}_\nu \hat{T} \hat{\gamma}_\mu == 2\hat{\eta}_{\mu\nu}. \tag{165}$$

The simpler case of GIR holds when the isometric coincides with a Riemannian metric, $\hat{\eta}_{\mu\nu} = g(x)_{\mu\nu}$, including the Schwartzschild metric, by therefore describing an isoelectron-isopositron pair under an external gravitational field.

The isogeometric unification of this section is illustrated by the fact that Eq.(163) allowed the representation of *all* characteristics of the neutron in its synthesis from the hydrogen with the sole change of the metric into one used for SIR, such as that of the EPR entanglement of Section 7.2.3 [70].

8.6.3. Black or brown holes? The study of IGIR has been rudimentarily initiated in Ref. [180] with intriguing outcome, such as the apparent *reformulation of black holes into brown holes* in representation of the lack of existence of singularities in nature.

Alternatively, the reformulation of black into brown holes appears to indicate the existence of a limit in the compressibility of protons at the divergence of their entangled number

8.6.4. Studies in the origin of gravitation. Recall that the gravitational field of a mass originates in the interior of the mass. Hence, a historical open problem is that of the *origin* rather than the sole description of a gravitational field.

It may be of some interest to know that IGIR can be defined as *a representation of the origin of the gravitational field with particular reference to the study of matter-matter gravitational attraction and matter-antimatter gravitational repulsion.*

As an example, a typical problem of the IGIR is the study of Poincaré's hypothesis that the exterior gravitation field of a mass is entirely generated by the electric and magnetic fields of its charged constituents.

The Poincaré hypothesis has been studied in detail in the 1974 MIT paper [185] via advanced and retarded formulations of quantum field theory. This study confirmed that the electromagnetic field of all charged components of a mass, including atomic and nuclear constituents, can indeed be the complete source of the exterior gravitational field of a mass even when the total charge is zero.

This study implies the existence of a gravitational source $\hat{F}^{elm}_{\mu\nu}$ representing the origin of the gravitational field, with ensuing completion of field equations (157) [180] into the form

$$\hat{R}_{\mu\nu} - \frac{1}{2}\hat{R}_{\mu\nu}\hat{R} = k\hat{F}^{elm}_{\mu\nu}, \tag{166}$$

which does verify the forgotten Freud identity of the Riemannian geometry [181].

Note that Eq. (163) no longer represent gravitation via curvature and this explains the reason for which the study of the Poincaré hypothesis in the origin of the gravitational field has been vastly ignored by mainstream physics for one full century (see debate [91] for details) despite Einstein's doubts on the r.h.s. of field equations (162) which he called *"a house made of wood"* while called the l.f.s. *"a house made of marble."*

For non-initiated readers, we should recall that Eq. (163) is fully admitted by conventional GR although, in this case, $\hat{F}^{elm}_{\mu\nu}$ represents the field of the *total charge* of the mass considered, with ensuing extremely small contribution to the gravitational field that, as such, remains represented by curvature.

By contrast, in field equation (163) according to the 1974 paper [185], the term $\hat{F}^{elm}_{44}$ represents the *total mass of the body assumed to be neutral,* the extension to the a non-null total charge being trivial. In this case, the gravitational field cannot be represented via curvature, because it is correctly represented by the iso-Minkowskian geometry that, as indicated earlier, is flat.

The basic open problems of the IGIR can then be tentatively formulated as the study of: the gravitational attraction of two masses represented with two equations of type (163)a; and the study of the gravitational repulsion between one mass represented with Eqs. (163) and its antimatter image represented via the isodual iso-Minkowskian geometry and isodual image of Eqs.m(163) [29] [70].

Additional studies in the general relativistic formulation of the electromagnetic field and its implications for quantum gravity are available from contributions [225] and [234] of the 2020 Teleconference [0].

8.6.5. Experimental test of the origin of gravitation. An experiment to prove or disprove the Poincaré hypothesis was suggested in page 145 of the 1974 MIT paper [185] (Figure 35). The U. S. *Hadronic Technologies Corporation* has constructed the equipment according proposed in Ref. [185] called *gravity generation equipment* (GGE), essentially consisting of a series of discs with *null* total charge of 1" thickness and diameter varying from 3" to 5".The discs are composed by various material, such as aluminum or Iron, which discs are put in rotation up to $100,000$ *rpm* by a specially designed sequence of spindles (Figure 34).

The proposed experiment essentially consists in: 1) placing the indicated GGE next to a highly sensitive gravity detector; 2) measuring the local gravitational field when the GGE is disconnected; and 3) measuring the local gravitational field when the GGE is progressively activated up to the maximal $100,000$ *rpm*. The detection of any gravitational field when the GGE discs are rotating would establish the first known *creation in laboratory of a gravitational field.* Note that there is no need for the GGE discs to be charged (see Ref.[185] for details).

In the author's view, Einstein's geometric conception of gravitation as being entirely due to curvature is one of the most beautiful and important discoveries by the human mind. However, such a conception has merely initiated, rather than ended, the studies in gravitation in view of

Figure 36: *It is generally accepted in the physics community that quantum mechanics is inapplicable in the interior of a black hole. Consequently, quantum mechanics should not be expected to be applicable in the interior of the scattering region of current high energy particle experiments in accordance with the EPR argument [1].*

century-old controversies yet to be resolved, and so much remains to be discovered theoretically and experimentally.

8.7. Application of EPR completions to high energy scattering experiments

The non-relativistic and relativistic elaboration of high energy scattering experiments with the EPR completion of the quantum into hadronic mechanics is known as:

i) *isoscattering theory,* when possessing a Lie-isotopic algebraic structure for the representation of *elastic,* thus time-reversal invariant scattering experiments on an iso-Hilbert isospace $\hat{\mathcal{H}}$ over the isofield $\hat{\mathcal{C}}$;

ii) *genoscattering theory,* when possessing a Lie-admissible algebraic structure for the representation of *inelastic,* thus time irreversible scattering experiments on an geno-Hilbert genospace $\hat{\mathcal{H}}^>$ over the genocomplex genofield $\hat{\mathcal{C}}^{>}$; and

iii) *hyperscattering theory,* when elaborated via hyperstructures for the representation of time irreversible multi-particle scattering experiments requiring a three-dimensional multi-valued representation on a hyper-Hilbert hyperspace over the hypercomplex hyperfield [24] [233].

Studies in the field were initiated by R. Mignani in 1984 [186] and continued by other authors (see the 1995 review in Chapter 12 of Ref. [24]). Subsequent detailed studies have been conducted by A. K. Aringazin *et al.* [54], A. O. E. Animalu *et al.* [187] [215].

Numerous additional works exist in the formulation of *non-unitary scattering theories,* but they are formulated on a conventional Hilbert space $\mathcal{H}$ over the conventional field of complex numbers $\mathcal{C}$, thus generally violating causality [23] [212]. Nevertheless the latter scattering theories are significant because they can be easily reformulated on the appropriate iso-, geno- or hyper-space over a corresponding iso-, geno- or hyperfield, by becoming iso-, geno- or hyper-unitary theories and regaining in this way causality.

By recalling the preservation of quantum mechanical axioms by hadronic mechanics (Section 8.4.4-I), and by remembering the dual formulation of hadronic mechanics on isospaces over isofields and their projection on conventional spaces over conventional fields, to the author's best knowledge the application of the EPR completions to high energy scattering experiments essentially implies the following:

a) The simplest possible Copenhagen realization of quantum mechanical axioms for point-particles in vacuum is no longer exactly valid for the hyperdense interior of high energy scattering regions due to their approaching the density of black holes, in favor of the broadest possible realization of quantum mechanical axions according to the iso-, geno- and hyper-structural branches of relativistic hadronic mechanics (Figure 36).

b) All theoretical as well as numerical results of high energy scattering experiments released by particle physics laboratories to date remain valid when formulated on an iso-, geno-, or hyper-Minkowskian isospace over an iso-, geno-, or hyper-complex fields.

c) The projection of the preceding formulations on our physical Minkowski space-time over the field of real numbers requires isorenormalization (133) of Isoaxiom IV of the energy and other isorenormalizations of the characteristics of intermediate particles which have been crucial for the exact representation of the neutron synthesis (Section 8.2.2), nuclear data (Section 8.2.5), molecular data (Section 7.2), and other data.

9. CONCLUDING REMARKS

The author has stated various times in his works that *the basic axioms of quantum mechanics and special relativity are majestic* in view of their mathematical consistency, predictive power, and preservation over time of numerical results for the conditions of their original conception (point particles in vacuum).

A central feature of the research outlined in this Overview is that the basic axioms of quantum mechanics are preserved identically in the verifications and applications [210]-[214] of the EPR argument [1], and they are merely realized in their most general possible form.

A similar situation occurs for special relativity because, when equally realized in their most general possible form, its axioms appear to allow: a geometric unification of Einstein's special and general relativities; their extension to interior dynamical problems; the apparent resolution of century-old controversies in gravitation; the study of the *origin* (rather than the sole description) of the gravitational field; the study of the mechanism of matter-matter gravitational attraction and matter-antimatter gravitational repulsion; and other intriguing open problems.

In closing, the EPR argument provides indeed grounds for the axiom-preserving completion of Copenhagen's simplest possible realization of quantum axioms into their broadest possible realization suggested by hadronic mechanics.

Acknowledgements.
The author would like to reinstate the rather long acknowledgements expressed in Refs. [36] [214], and additionally thank all participants of the 2020 Teleconference [0] for penetrating criticisms and comments. Special thanks are due to Carla Santilli, Founder and President of *Hadronic Press, Inc.*, for her role in recording the laborious efforts over the past half a century in the construction of hadronic mechanics and chemistry and making the publications available in free pdf download for this Overview. Special thanks are finally due to Erik Trell for a detailed technical control and to Sherri Stone for an accurate linguistic control of the manuscript. Needless to say, the author is solely responsible for the content of this Overview due to numerous revisions done in the final version.

REFERENCES

[0] International teleconference on Einstein-Podolsky-Rosen argument that "Quantum mechanics is not a complete theory,", September 1 to 5, 2020,
Announcement
http://eprdebates.org/epr-conference-2020.php
Recording of lectures and comments
http://www.world-lecture-series.org/level-xii-epr-teleconference-2020

[1] A. Einstein, B. Podolsky and N. Rosen, "Can quantum-mechanical description of physical reality be considered complete?," Phys. Rev. **47**, 777 (1935),
http://www.eprdebates.org/docs/epr-argument.pdf

[2] N. Bohr, "Can quantum mechanical description of physical reality be considered complete?" Phys. Rev. **48**, 696 (1935),
www.informationphilosopher.com/solutions/scientists/bohr/EPRBohr.pdf

[3] J. S. Bell: "On the Einstein Podolsky Rosen paradox" Physics **1**, 195 (1964),
https://cds.cern.ch/record/111654/files/vol1p195-200_001.pdf

[4] Stanford Encyclopedia of Philosophy, "Bell's Theorem" (2019),
https://plato.stanford.edu/entries/bell-theorem

[5] Sudhakar S. Dhondge "Studies on Santilli Three-Body Model of the Deuteron According to Hadronic Mechanics," American Journal of Modern Physics **5**, 46-55 (2016),
http://www.santilli-foundation.org/docs/deuteron-2018.pdf

[6] R. M. Santilli, "Embedding of Lie-algebras into Lie-admissible algebras," Nuovo Cimento **51**, 570 (1967),
http://www.santilli-foundation.org/docs/Santilli-54.pdf

[7] R. M. Santilli, "Dissipativity and Lie-admissible algebras," Meccanica **1**, 3 (1969).

[8] R. M. Santilli, "An introduction to Lie-admissible algebras," Suppl. Nuovo Cimento, **6**, 1225 (1968).

[9] A. A. Albert, Trans. Amer. Math. Soc. **64,** 552 (1948).

[10] R. M. Santilli, "Representation of antiparticles via isodual numbers, spaces and geometries," Comm. Theor. Phys. **3,** 153 (1994),
http://www.santilli-foundation.org/docs/Santilli-112.pdf

[11] P. A. M. Dirac, Proceedings of the Royal Society, **117**, 610 (1928).

[12] Vijay M. Tange, "Advances in hadronic chemistry and its applications," Foundation of Chemistry, DOI 10.1007/s10698-015-9218-z (2015),
http://www.santilli-foundation.org/docs/hadronic-chemistry-FC.pdf

[13] G. Eigen, "Measurements of polarization correlation of the two-photon system produced in positron-electron annihilation," Lecture delivered at the 2020 Teleconference [0],
http://www.world-lecture-series.org/eigen-measurements-of-polarization-correlation-of-the-two-photon-system-produced-in-positron-electron-annihilation

[14] M. Fadel, T. Zibold, B. Decamps, Ph. Treutlein, "Spatial entanglement pat-

terns and Einstein-Podolsky-Rosen steering in Bose-Einstein condensates, " Science **360**, 409 (2018),
http://www.santilli-foundation.org/Basel-paper.pdf

[15] J. Schukraft, Heavy-ion physics with the ALICE experiment at the CERN Large Hadron Collider,
https://royalsocietypublishing.org/doi/10.1098/rsta.2011.0469

[16] W. Heisenberg, Nachr. Akad. Wiss. Gottingen **IIa**, 111 (1953),
https://link.springer.com/chapter/10.1007/978-3-642-70079-8_23

[17] "Bohmian (de Broglie-Bohm) Mechanics," Stanford Encyclopedia of Philosophy,
https://plato.stanford.edu/entries/qm-bohm/

[18] D. Bohm, "A Suggested Interpretation of the Quantum Theory in Terms of "Hidden Variables," Physical Review, **85**, 166, (1952),
https://journals.aps.org/pr/abstract/10.1103/PhysRev.85.166

[19] R. M. Santilli, "On a possible Lie-admissible covering of Galilei's relativity in Newtonian mechanics for nonconservative and Galilei form-non-invariant systems," Hadronic J. **1**, 223 (1978),
http://www.santilli-foundation.org/docs/Santilli-58.pdf

[20] R. M. Santilli, "Need of subjecting to an experimental verification the validity within a hadron of Einstein special relativity and Pauli exclusion principle," Hadronic J. **1**, pages 574 (1978),
http://www.santilli-foundation.org/docs/santilli-73.pdf

[21] R. M. Santilli, *Foundation of Theoretical Mechanics,* Springer-Verlag, Heidelberg, Germany, Volume I (1978) *The Inverse Problem in Newtonian Mechanics,*
http://www.santilli-foundation.org/docs/Santilli-209.pdf

[22] R. M. Santilli, *Foundation of Theoretical Mechanics,* Springer-Verlag, Heidelberg, Germany, Vol. II (1982) *Birkhoffian Generalization of Hamiltonian Mechanics,*

http://www.santilli-foundation.org/docs/santilli-69.pdf

[23] R. M. Santilli, *Elements of Hadronic Mechanics*, Ukraine Academy of Sciences, Kiev, Volume I (1995), *fMathematical Foundations*, http://www.santilli-foundation.org/docs/Santilli-300.pdf

[24] R. M. Santilli, *Elements of Hadronic Mechanics*, Ukraine Academy of Sciences, Kiev, Volume II (1994), *Theoretical Foundations*, www.santilli-foundation.org/docs/Santilli-301.pdf

[25] R. M. Santilli, *Elements of Hadronic Mechanics*, Ukraine Academy of Sciences, Kiev, Volume III (2016), *Experimental verifications*, www.santilli-foundation.org/docs/elements-hadronic-mechanics-iii.compressed.pdf

[26] R. M. Santilli, *The Physics of New Clean Energies and Fuels According to Hadronic Mechanics*, Special issue of the Journal of New Energy, 318 pages (1998),
http://www.santilli-foundation.org/docs/Santilli-114.pdf

[27] R. M. Santilli, *Isotopic, Genotopic and Hyperstructural Methods in Theoretical Biology*, Ukraine Academy of Sciences, Kiev (1994),
http://www.santilli-foundation.org/docs/santilli-67.pdf

[28] B. Davvaz and Thomas Vougiouklis, *A Walk Through Weak Hyperstructures, Hv-Structures*, World Scientific (2018).

[29] R. M. Santilli, *Isodual Theory of Antimatter with Applications to Antigravity, Grand Unifications and Cosmology*, Springer (2006),
www.santilli-foundation.org/docs/santilli-79.pdf

[30] R. M. Santilli, *Foundations of Hadronic Chemistry, with Applications to New Clean Energies and Fuels*, Kluwer Academic Publishers (2001),
http://www.santilli-foundation.org/docs/Santilli-113.pdf
Russian translation by A. K. Aringazin,
http://i-b-r.org/docs/Santilli-Hadronic-Chemistry.pdf

[31] R. M. Santilli and D. D. Shillady, "A new isochemical model of the hydrogen molecule," Intern. J. Hydrogen Energy **24**, 943 (1999),
http://www.santilli- foundation.org/docs/Santilli-135.pdf

[32] R. M. Santilli and D. D. Shillady, "A new isochemical model of the water molecule," Intern. J. Hydrogen Energy **25**, 173 (2000),
http://www.santilli-foundation.org/docs/Santilli-39.pdf

[33] R. M. Santilli, "Isonumbers and Genonumbers of Dimensions 1, 2, 4, 8, their Isoduals and Pseudoduals, and "Hidden Numbers," of Dimension 3, 5, 6, 7," Algebras, Groups and Geometries **10**, 273 (1993),
http://www.santilli-foundation.org/docs/Santilli-34.pdf

[34] R. M. Santilli, "Nonlocal-Integral Isotopies of Differential Calculus, Mechanics and Geometries," in Isotopies of Contemporary Mathematical Structures," Rendiconti Circolo Matematico Palermo, Suppl. **42**, 7 (1996),
http://www.santilli-foundation.org/docs/Santilli-37.pdf

[35] R. M. Santilli, "Invariant Lie-isotopic and Lie-admissible formulation of quantum deformations," Found. Phys. **27**, 1159 (1997),
http://www.santilli-foundation.org/docs/Santilli-06.pdf

[36] R. M. Santilli, "Relativistic hadronic mechanics: nonunitary, axiom-preserving completion of relativistic quantum mechanics," Found. Phys. **27**, 625 (1997),
http://www.santilli-foundation.org/docs/Santilli-15.pdf

[37] R. M. Santilli, "Initiation of the representation theory of Lie-admissible algebras of operators on bimodular Hilbert spaces," Hadronic J. **3**, 440 (1979),
http://www.santilli-foundation.org/docs/santilli-1978-paper.pdf

[38] R. M. Santilli, "An intriguing legacy of Einstein, Fermi, Jordan and others: The possible invalidation of quark conjectures," Found. Phys. **11**, 384 (1981),
www.santilli-foundation.org/docs/Santilli-36.pdf

[39] R. M. Santilli, *Lie-Admissible Approach to the Hadronic Structure,* Vol. I *Non-Applicability of the Galileo and Einstein Relativities?* Hadronic Press (1982),

http://www.santilli-foundation.org/docs/santilli-71.pdf
Vol. II *Covering of the Galileo and Einstein Relativities?* Hadronic Press (1982),
http://www.santilli-foundation.org/docs/santilli-72.pdf

[40] R. M. Santilli, "A quantitative isotopic representation of the deuteron magnetic moment," in *Proceedings of the International Symposium 'Dubna Deuteron-93,* Joint Institute for Nuclear Research, Dubna, Russia (1994),
http://www.santilli-foundation.org/docs/Santilli-134.pdf

[41] R. M. Santilli, "Lie-admissible invariant representation of irreversibility for matter and antimatter at the classical and operator levels," Nuovo Cimento B **121**, 443 (2006),
http://www.i-b-r.org/Lie-admiss-NCB-I.pdf

[42] J. Fronteau, A. Tellez-Arenas and R. M. Santilli, "Lie-admissible structure of statistical mechanics," Hadronic J. **3**, 130 (1979),
http://www.santilli-foundation.org/docs/arenas-fronteau-santilli-1981.pdf

[43] H. Rauch and A. Zeilinger, "Demonstration of SU(2) symmetry by neutron interferometers," in the *Proceedings of the 1981 Third Workshop on Lie-Admissible Formulations,* Hadronic J., **4**, 1280 (1981),
http://www.santilli-foundation.org/docs/ rauch-zeilinger-1981.pdf

[44] C. Eder, "On the mutation parameters of the generalized spin algebra with spin 1/2,"n the *Proceedings of the 1981 Third Workshop on Lie-Admissible Formulations,* Hadronic J., **4**, 634 (1981),
http://www.santilli-foundation.org/docs/eder-1981.pdf

[45] H. C. Myung and R. M. Santilli, "Modular-isotopic Hilbert space formulation of the exterior strong problem," Hadronic Journal **5**, 1277 (1982),
http://www.santilli-foundation.org/docs/myung-santilli-1982.pdf

[46] A. O. E. Animalu and R. M. Santilli, in *Hadronic Mechanics and Nonpotential Interactions,* M. Mijatovic, Editor, Nova Science, New York, p. 19-22 (1990).

[47] R. M. Santilli, "Generalization of Heisenberg's uncertainty principle for

strong interactions," Hadronic Journal **4**, 642 (1981),
http://www.santilli-foundation.org/docs/generalized-uncertainties-1981.pdf

[48] J. V. Kadeisvili, "Elements of functional isoanalysis," Algebras, Groups and Geometries **9**, 283 (1992).

[49] J. V. Kadeisvili, "Elements of the Fourier-Santilli isotransforms," Algebras, Groups and Geometries **9**, 319 (1992).

[50] J. V. Kadeisvili, '"An introduction to the Lie-Santilli isotopic theory," Mathematical Methods in Applied Sciences **19**, 134 (1996),
http://www.santilli-foundation.org/docs/Santilli-30.pdf

[51] A. K. Aringazin, D. A. Kirukhin, and R. M. Santilli, "Isotopic Generalization of the Legendre, Jacobi, and Bessel Functions," Algebras, Groups and Geometries **12**, 255 (1995).

[52] J. Dunning-Davies, The Thermodynamics Associated with Santilli's Hadronic Mechanics Progress in Physics, **4**, 24 (2006),
http://www.santilli-foundation.org/docs/Dunning-Davies-Thermod.PDF

[53] A. A. Bhalekar, "Santilli's Lie-Admissible Mechanics. The Only Option Commensurate with Irreversibility and Nonequilibrium Thermodynamics," AIP Conference Proceedings 1558, 702 (2013);
doi: 10.1063/1.4825588,
www.santilli-foundation.org/docs/bhalekar-lie-admissible.pdf

[54] A. K. Aringazin, D.A. Kirukhin, and R.M. Santilli, "Nonpotential Elastic Scattering of Spinning Particles," Part I, Hadronic J. **18**, 245 (1995),
http://www.santilli-foundation.org/docs/aringazin-part-1.pdf
Part II, Hadronic J. **18**, 257 (1995):
http://www.santilli-foundation.org/docs/aringazin-part-2.pdf

[55] A. S. Muktibodh and R. M. Santilli, "Studies of the Regular and Irregular Isorepresentations of the Lie-Santilli Isotheory," Journal of Generalized Lie Theories, **11**, 1, (2007),

http://www.santilli-foundation.org/docs/isorep-Lie-Santilli-2017.pdf

[56] A. O. E. Animalu and R. M. Santilli, "Nonlocal isotopic representation of the Cooper pair in superconductivity," Intern. J. Quantum Chemistry **29**, 185 (1995),
http://www.santilli-foundation.org/docs/Santilli-26.pdf

[57] A. S Muktibodh, "Isodual theory of antimatter, I: Mathematical and physical methods," Algebras, Groups and Geometries **35**, 67 (2018),
http://www.santilli-foundation.org/docs/ Muktibodh-agg-isoduasl.pdf

[58] R. M. Santilli, "Does antimatter emit a new light?" Invited paper for the proceedings of the International Conference on Antimatter, held in Sepino, Italy, on May 1996, published in Hyperfine Interactions **109**, 63 (1997),
http://www.santilli-foundation.org/docs/Santilli-28.pdf

[59] A. Schoeber, Editor, *Collected Papers on Irreversibility and Non-potentiality in Statistical Mechanics,* Hadronic Press (1984),
http://www.santilli-foundation.org/docs/Santilli-110.pdf

[60] R. M. Santilli, "Lie-isotopic Lifting of Special Relativity for Extended Deformable Particles," Lettere Nuovo Cimento**37**, 545-555 (1983),
www.santilli-foundation.org/docs/Santilli-50.pdf

[61] R. M. Santilli, "Lie-isotopic Lifting of Unitary Symmetries and of Wigner's Theorem for Extended and Deformable Particles," Lettere Nuovo Cimento **38**, 509 (1983),
www.santilli-foundation.org/docs/Santilli-51.pdf

[62] R. M. Santilli, "Isotopies of Lie Symmetries," I: Basic theory," Hadronic J. **8**, 8 (985),
www.santilli-foundation.org/docs/santilli-65.pdf

[63] R. M. Santilli, "Isotopies of Lie Symmetries," II: Isotopies of the rotational symmetry," Hadronic J. **8**, 36 (1985),
www.santilli-foundation.org/docs/santilli-65.pdf

[64] R. M. Santilli, "Rotational isotopic symmetries," ICTP communication No. IC/91/261 (1991),
www.santilli-foundation.org/docs/Santilli-148.pdf

[65] R. M. Santilli, "Isotopic Lifting of the SU(2) Symmetry with Applications to Nuclear Physics," JINR rapid Comm. **6**. 24 (1993),
www.santilli-foundation.org/docs/Santilli-19.pdf

[66] R. M. Santilli, "Isotopic generalization of the Poincare' symmetry," ICTP communication No. IC/91/45 (1991),
www.santilli-foundation.org/docs/Santilli-140.pdf

[67] R. M. Santilli, "Nonlinear, Nonlocal and Noncanonical Isotopies of the Poincare' Symmetry," Moscow Phys. Soc. **3**, 255 (1993),
http://www.santilli-foundation.org/docs/Santilli-40.pdf

[68] R. M. Santilli, "Isotopies of the spinorial covering of the Poincaré symmetry," Communication of the Joint Institute for Nuclear Research, Dubna, Russia, No. E4-93-252 (1993).

[69] R. M. Santilli, Chinese J. System Eng. and Electr. **6**, 177 (1995),
www.santilli-foundation.org/docs/Santilli-18.pdf

[70] R. M. Santilli, "Isominkowskian Geometry for the Gravitational Treatment of Matter and its Isodual for Antimatter," Intern. J. Modern Phys. D **7**, 351 (1998),
www.santilli-foundation.org/docs/Santilli-35.pdf

[71] R. M. Santilli, *Isotopic Generalizations of Galilei and Einstein Relativities*, International Academic Press (1991), Vols. I *Mathematical Foundations*,
http://www.santilli-foundation.org/docs/Santilli-01.pdf

[72] R. M. Santilli, *Isotopic Generalizations of Galilei and Einstein Relativities*, International Academic Press (1991), Vol. II: *Classical Formulations*,
www.santilli-foundation.org/docs/Santilli-61.pdf

[73] R. M. Santilli, *Isorelativities,* International Academic Press, (1995), new edition to appear.

[74] R. M. Santilli, *Hadronic Mathematics, Mechanics and Chemistry,*, Volumes I to V, International Academic Press, (2008),
www.i-b-r.org/Hadronic-Mechanics.htm

[75] R. M. Santilli, "Does antimatter emit a new light?" Invited paper for the proceedings of the International Conference on Antimatter, held in Sepino, Italy, on May 1996, published in Hyperfine Interactions, **109**, 63 (1997),
http://www.santilli-foundation.org/docs/Santilli-28.pdf

[76] EPR debate on cosmology,
http://eprdebates.org/no-universe-expansion.php

[77] F. Zwicky, "On the Red Shift of Spectral Lines through Interstellar Space," Proceedings of the National Academy of Sciences of the United States of America, **15**, 773 (1929),
http://www.pnas.org/cgi/reprintframed/15/10/773

[78] H. Ahmar, G. Amato, J. V. Kadeisvili, J. Manuel, G. West, and O. Zogorodnia, "Additional experimental confirmations of Santilli's IsoRedShift and the consequential lack of expansion of the universe,"Journal of Computational Methods in Sciences and Engineering, **13**, 321 (2013),
http://www.santilli-foundation.org/docs/IRS-confirmations-212.pdf

[79] R. M. Santilli, "Representation of galactic dynamics via isoshifts without universe expansion, dark matter and dark energy," American Journal of Modern Physics **4**, 26, (2015),
http://www.hadronictechnologies.com/docs/dark-matter-2015.pdf

[80] P. A. Laviolette "Is the universe really expanding?" The Astrophysical Journal **301**, 544 (1986).

[81] H. Arp. *Quasars Redshift and Controversies.* Interstellar Media, Berkeley

(1987).

[82] R. M. Santilli, "Apparent Detection of a New Antimatter Galaxy in the Capella Region of the Night Sky," Clifford Analysis, Clifford Algebras and their Applications, in press (2016),
http://www.santilli-foundation.org/docs/capella-antimatter-galaxy.pdf

[83] S. Beghella-Bartoli, R. M. Santilli, "Possible Role of Antimatter Galaxies for the Stability of the Universe," American Journal of Modern Physics 5(2-1): 185-190 (2016),
http://www.santilli-foundation.org/docs/pdf11.pdf

[84] R. M. Santilli, "Apparent consistency of Rutherford's hypothesis on the neutron as a compressed hydrogen atom, Hadronic J. **13**, 513 (1990),
http://www.santilli-foundation.org/docs/Santilli-21.pdf

[85] R. M. Santilli,"Nonlocal formulation of the Bose-Einstein correlation within the context of hadronic mechanics," Hadronic J. **15**, 1 and 81 (1992),
www.santilli-foundation.org/docs/Santilli-116.pdf

[86] F. Cardone and R. Mignani,"Metric description of hadronic interactions from the Bose-Einstein correlation,"
JETP **83**, 435 (1996),
www.santilli-foundation.org/docs/Santilli-130.pdf

[87] J. V. Kadeisvili, "The Rutherford-Santilli Neutron," Hadronic J. **31**, 1 (2008),
http://www.i-b-r.org/Rutherford-Santilli-II.pdf
also available in html format at
http://www.i-b-r.org/Rutherford-Santilli-neutron.htm

[88] E. Rutherford, Proc. Roy. Soc. A **97**, 374 (1920).

[89] R. M. Santilli, "The etherino and/or the Neutrino Hypothesis?" Found. Phys. **37**, 670 (2007),
www.santilli-foundation.org/docs/EtherinoFoundPhys.pdf

[90] R. M. Santilli, "Perché lo spazio é rigido" (Why space is rigid), Il Pungolo Verde, Campobasso, Italy (1956),
Italian version
http://www.santilli-foundation.org/docs/rms-56.pdf
English translation
http://www.santilli-foundation.org/docs/rms-56-english.pdf

[91] EPR debate "Etherino or neutrino,"
http://eprdebates.org/etherino-neutrino.php

[92] R. Norman and J. Dunning-Davies, "Hadronic paradigm assessed: neutroid and neutron synthesis from an arc of current in hydrogen gas," Hadronic Journal **40**, 119 (2017),
http://santilli-foundation.org/docs/norman-dunning-davies-hj.pdf

[93] R. M. Santilli, "Apparent confirmation of Don Borghi's experiment on the laboratory synthesis of neutrons from protons and electrons, Hadronic J. **30**, 29032 (2007),
http://www.i-b-r.org/NeutronSynthesis.pdf

[94] R. Norman, S. Beghella Bartoli, B. Buckley, J. Dunning-Davies, J. Rak, R. M. Santilli "Experimental Confirmation of the Synthesis of Neutrons and Neutroids from a Hydrogen Gas, American Journal of Modern Physics, **6**, 85 (2017),
http://www.santilli-foundation.org/docs/confirmation-neutron-synthesis-2017.pdf

[95] A. A. Bhalekar and R. M. Santilli, "Exact and Invariant representation of nuclear magnetic moments and spins according to hadronic mechanics," American Journal of Modern Physics **5**, 56 (2016),
http://www.santilli-foundation.org/docs/nuclear-MM-spins.pdf

[96] R. M. Santilli, "Apparent Experimental Confirmation of Pseudoprotons and their Application to New Clean Nuclear Energies," International Journal of Applied Physics and Mathematics Volume 9, Number 2 (2019),
www.santilli-foundation.org/docs/pseudoproton-verification-2018.pdf

[97] R. M. Santilli, "Hadronic energy," Hadronic J. **17**, 311 (1994),

http://www.santilli-foundation.org/docs/hadronic-energy.pdf
"Conceivable new recycling of nuclear waste by nuclear power companies in their plants, " arXiv:physics/9704014 (1997),
https://arxiv.org/pdf/physics/9704014.pdf

[98] U. Abundo, "Interpretation and enhancement of the excess energy of Rossi's reactor via Santilli neutroids and nucleoids," Hadronic Journal **37**, 697 (2014),
http://www.hadronictechnologies.com/docs/abundo-paper-2014.pdf

[99] Nuclear Data Center at KAERI Table of Nuclides,
http://atom.kaeri.re.kr:81
http://atom.kaeri.re.kr:81/ton/nuc1.html

[100] N. F. Tsagas, A. Mystakidis, G. Bakos, L. Sfetelis, D. Koukoulis, and S. Trassanidis, " Experimental verification of Santilli's clean subnuclear hadronic energy," Hadronic Journal **19**, 87 (1996),
http://www.santilli-foundation.org/docs/N-Tsagas-1996.pdf

[101] R. M. Santilli, "Invariant Lie-Admissible Classical and Operator Mechanics for Matter, Lecture,
http://www.world-lecture-series.org/level-iii

[102] R. M. Santilli, "The novel "Controlled Intermediate Nuclear Fusion:" A report on its industrial realization as predicted by hadronic mechanics," Hadronic J. **31**, 1 (2008),
http://www.i-b-r.org/CNF-printed.pdf

[103] R. M. Santilli, "Experimental Confirmation of Nitrogen Synthesis from deuterium and Carbon without harmful radiations," New Advances in Physics **4**, Issue no.1, (2010),
http://www.santilli-foundation.org/docs/Nitrogen-synthesis-2010.pdf

[104] D. Rossiter, Director, "IVA Report 184445 on comparative Nitrogen counts on samples of the Nitrogen synthesis,"
www.santilli-foundation.org/docs/Spectral-analysis-Ref-[79].png

[105] D. Rossiter, Director, "IVA Report 189920 on comparative Silica counts," www.santilli-foundation.org/docs/IVAReport 189920.pdf

[106] D. Rossiter, Director, "IVA Report 189920 on comparative Silica count," www.santilli-foundation.org/docs/IVAReport 189920.pdf

[107] D. Rossiter, Director, "IVA Report 200010 on comparative Nitrogen counts, http://www.santilli-foundation.org/docs/Oneida-analyses-2013.zip

[108] R. M. Santilli, "Sound of Intermediate Controlled nuclear syntheses," www.santilli-foundation.org/Thunder-Fusions.amr

[109] R. Brenna, T. Kuliczkowski, L. Ying, "Verification of Santilli intermediate Controlled Nuclear Fusions without harmful radiations and the production of magnecular clusters," New Advances in Physics, **5**, 9 (2011), http://www.santilli-foundation.org/docs/ICNF-2.pdf

[110] R. Brenna, T. Kuliczkowski and L. Ying, "Report on Test for Silicon" on the Nitrogen synthesis, http://www.santilli-foundation.org/docs/PGTI-Anal-test1.pdf

[111] L. Ying, W. Cai, J. , C. Lynch, S. Marton, S. Elliot and Y. Yang, "Experimental verification for Intermediate Controlled Nuclear Fusion," City College of New York, Preprint (2012), unpublished, http://www.santilli-foundation.org/docs/ICNF-Cai-paper- Ying.pdf

[112] Leong Ying, "Verification of Santilli's Intermediate Nuclear Harmful Radiation and the Production of Magnecular Clusters," Lecture (2012), http://www.world-lecture-series.org/lecture-vd

[113] R. M. Santilli, "Video presentation of the third hadronic reactor for the Nitrogen and Silicon syntheses," (2011), http://www.world-lecture-series.org/dragon-iii

[114] R. M. Santilli, "Additional Confirmation of the Intermediate Controlled Nuclear Fusions without harmful radiation or waste" in the *Proceedings of the*

Third International Conference on the Lie-Admissible Treatment of Irreversible Processes, Kathmandu University (2011) pages 163-177,
http://www.santilli-foundation.org/docs/ICNF-3.pdf

[115] R. M. Santilli, “Intermediate Controlled Nuclear Fusions without the emission of radiations and without the release of radioactive waste,” Lecture,
http://www.world-lecture-series.org/level-v

[116] D. Swartz, "Constellation Technologies first report on comparative Silicon counts,"
http://www.santilli-foundation.org/docs/Constellation-Si-10-13.zip

[117] D. Swartz, "Constellation technologies second report on comparative Silicon counts,"
http://www.santilli-foundation.org/docs/Constellation-Rep-Si-2.zip

[118] D. Swartz, "Constellation technologies third report on comparative Silicon counts,"
http://www.santilli-foundation.org/docs/Constell-Si-3.pdf

[119] A. Nas, "Data on Constellation technologies tests 1 and 2 on comparative Silicon counts,"
http://www.santilli-foundation.org/docs/Data-Constell-tests.docx

[120] D. Swartz, "Constellation technologies Third report on comparative Silicon counts,"
http://www.santilli-foundation.org/docs/Constell-Silicon-10-14.pdf

[121] N. F. Tsagas, "Experimental attempt to verify Santilli's clean, subnuclear hadronic energy", IARD Lecture 2008,
http://www.santilli-foundation.org/docs/N-Tsagas-neutron-decay.pdf

[122] V. Tandge, "Hadronic Chemistry Applied to Hydrogen and Water Molecules," AIP Conf. Proc. 1479, 1013 (2012),
http://www.santilli-foundation.org/docs/Tandge-AIP.pdf
A. A. Bhalekar, R. M. Santilli, “Two Body IsoElectronium Model of the Heliunic

System," American Journal of Modern Physics **6**, 29 (2017),
http://www.santilli-foundation.org/docs/bhalekar-santilli-isohelium.pdf

[123] R. M. Santilli, "Method and Apparatus for the industrial production of new hydrogen-rich fuels," United States Patent Number 9,700,870, B2, July 1 (2017),
http://www.santilli-foundation.org/docs/Magnecule-patent.pdf

[124] R. M. Santilli, *The New Fuels with Magnecular Structure,* International Academic Press (2008),
http://www.i-b-r.org/docs/Fuels-Magnecular-Structure.pdf
Italian translation by G. Bonfanti
http://www.santilli-foundation.org/docs/Carb-Strutt-Magnecolare.pdf

[125] A. K. Aringazin and R. M. Santilli, "Structure and Combustion of Magnegases," arXiv:physics/0112066 (2001)
https://arxiv.org/pdf/physics/0112066.pdf
"A Study of the Energy Efficiency of Hadronic Reactors of Molecular Type," arXiv:physics/0112067 (2001)
https://arxiv.org/pdf/physics/0112067.pdf
A Study of Polycarbonyl Compounds in Magnegases," arXiv:physics/0112068 (2001),
https://arxiv.org/pdf/physics/0112068.pdf

[126] R. M. Santilli, "The Novel Sustainable Hyper-Combustion and Hyper-Furnaces of Thunder Energies Corporation," 2016 International Summit on the Environment, Hainan Island, China,
https://www.youtube.com/watch?v=eyS6HDO5xDc&feature=youtu.be

[127] Y. Yang, J. V. Kadeisvili, and S. Marton, "Experimental Confirmations of the New Chemical Species of Santilli Magnecules," The Open Physical Chemistry Journal **5**, 1 (2013),
www.santilli-foundation.org/docs/Magnecules-2012.pdf

[128] C. P. Pandhurneka and Sangesh P. Zodape, "Santilli's Magnecules and Their Applications," American Journal of Modern Physics); 6(4-1): 64-773 (2017),

http://www.santilli-foundation.org/docs/magnecules-2017.pdf

[129] R. R. Alfano, "CCNY Certification of Magnegas Flame Temperature," Summary,
http://hadronictechnologies.com/docs/Magnegas-flame

[130] R. R. Alfano, "CCNY Certification of Magnegas Flame Temperature," Report,
http://hadronictechnologies.com/docs/MG-Flame-report.pdf

[131] R. M. Santilli, "An Introduction to Hadronic Chemistry," Invited Lecture delivered at the Institute for UltraFast Spectroscopy and Laser City College of New York (2012),
http://www.world-lecture-series.org/lecture-iv-a

[132] N. Kapustka, "EMV Evaluation of Oxyfuel Gas Cutting Gases,"
http://hadronictechnologies.com/docs/MG-metal-cutting.pdf

[133] R. M. Santilli, "The novel magnecular species of hydrogen and oxygen with increased specific weight and energy content," Intern. J. Hydrogen Energy **28**, 177 (2003),
http://www.santilli-foundation.org/docs/Santilli-38.pdf

[134] D. Day, TCD analysis and density measurements of Santilli Magnehydrogen. Eprida Laboratory report dated 11/10/2011
www.santilli-foundation.org/docs/Eprida-MH-Certification-10-11.pdf

[135] S. P. Zodape, "The MagneHydrogen in Hadronic Chemistry," This work is being presented at ICNAAM 2013 being held at Rhodes, Greece during or. AIP Proceedings 1558, 648; doi: 10.1063/1.4825575 (2014),
www.santilli-foundation.org/docs/sangesh-Greece.pdf

[136] Y. Yang, J. V. Kadeisvili, and S. Marton, "Experimental Confirmations of the New Chemical Species of Santilli MagneHydrogen," International Journal Hydrogen Energy **38**, 5002 (2013),
http://www.santilli-foundation.org/docs/MagneHydrogen-2012.pdf

[137] C. P. Pandhurnekar, "Advances on Alternative Fuels with Santilli Magnecular Structure," International Journal of Alternative Fuels, **17**, 23 (2015), http://www.santilli-foundation.org/docs/Magnegas-2015.pdf

[138] R. M. Santilli "A Tentative Magnecular Model of Liquid Water with an Explicit Attractive Force Between Water Molecules," American Journal of Modern Physics **6**, 46 (2017),
www.santilli-foundation.org/docs/santilli-liquid-water.pdf

[139] R. M. Santilli, "A new gaseous and combustible form of water," Intern. J. Hydrogen Energy **31**, 1113 (2006),
http://www.santilli-foundation.org/docs/Santilli-138.pdf

[140] R. M. Santilli, "The Novel Hyper Combustion for the Complete Combustion of Fossil Fuels", Intern. Journal of Chemical Engineering and Applications, **10**, 16 (2019),
http://www.santilli-foundation.org/docs/hypercombustion-2019.pdf

[141] R. M. Santilli, "An Introduction to Hadronic Chemistry," Invited Lecture at the Institute for UltraFast Spectroscopy and Laser City College of New York (2012),
http://www.world-lecture-series.org/level-iv

[142] V. Tandge, "An introduction to hadronic chemistry," Invited Lecture, IC-NAAM (2012),
https://www.youtube.com/watch?v=NqBrFVVYWLc

[143] R. M. Santilli, Tutoring Lecture I: Isomathematics,
http://www.world-lecture-series.org/santilli-tutoring-i

[144] R. M. Santilli, Tutoring Lecture II: Verifications of the EPR argument,
http://www.world-lecture-series.org/santilli-tutoring-ii

[145] R. M. Santilli, Tutoring Lecture IV: Lie-admissible formulations,
http://www.world-lecture-series.org/santilli-tutoring-iv-part-1
http://www.world-lecture-series.org/santilli-tutoring-iv-part-2

http://www.world-lecture-series.org/santilli-tutoring-iv-part-3

[146] T. Vougiouklis, Tutoring Lecture V: Hypermathematics, http://www.world-lecture-series.org/vougiouklis-extending-mathematical-methods-from-numbers-to-hypernumbers-and-to-h_v-numbers

[147] R. M. Santilli, "Closed systems with non-Hamiltonian internal forces," ICTP release IC/91/259 (1991), www.santilli-foundation.org/docs/Santilli-143.pdf

[148] R. M. Santilli, "Inequivalence of exterior and interior dynamical problems" ICTP release IC/91/258 (1991), www.santilli-foundation.org/docs/Santilli-142.pdf

[149] R. M. Santilli, "Generalized two-body and three-body systems with non-Hamiltonian internal forces-" ICTP release IC/91/260 (1991), www.santilli-foundation.org/docs/Santilli-139.pdf

[150] R. M. Santilli, "Galileo-isotopic symmetries," ICTP release IC/91/263 (1991), www.santilli-foundation.org/docs/Santilli-147.pdf

[151] R. M. Santilli, " Galileo-Isotopic relativities," ICTP release (1991), www.santilli-foundation.org/docs/Santilli-146.pdf

[152] R. M. Santilli, "Theory of mutation of elementary particles and its application to Rauch's experiment on the spinorial symmetry," ICTP release IC/91/46 (1991), www.santilli-foundation.org/-docs/Santilli-141.pdf

[153] R. M. Santilli, "The notion of non-relativistic isoparticle," ICTP release IC/91/265 (1991), www.santilli-foundation.org/docs/Santilli-145.pdf

[154] A. K. Aringazin, A. Jannussis, F. Lopez, M. Nishioka and B. Vel-janosky, *Santilli's Lie-Isotopic Generalization of Galilei and Einstein Relativities*, Kostakaris Publishers, Athens, Greece (1991),

www.santilli-foundation.org/docs/Santilli-108.pdf

[155] R. M. Santilli, "Experimental Verifications of IsoRedShift with Possible Absence of Universe Expansion, Big Bang, Dark Matter, and Dark Energy," The Open Astronomy Journal **3**, 124 (2010),
http://www.santilli-foundation.org/docs/Isoredshift-Letter.pdf

[156] R. M. Santilli, "Experimental Verification of IsoRedShift and its Cosmological Implications," AIP Conference Proceedings Vol. 1281, pp. 882-885 (2010),
http://www.santilli-foundation.org/docs/Santilli-isoredshift.pdf

[157] R. M. Santilli, G. West and G. Amato, "Experimental Confirmation of the IsoRedShift at Sunset and Sunrise with Consequential Absence of Universe Expansion and Related Conjectures," Journal of Computational Methods in Sciences and Engineering, **12**, 165 (2012),
http://www.santilli-foundation.org/docs/Confirmation-sun-IRS.pdf

[158] G. West and G. Amato, "Experimental Confirmation of Santilli's IsoRedShift and IsoBlueShift," lecture delivered at the *San Marino Workshop on Astrophysics and Cosmology for Matter and Antimatter*, Republic of San Marino (2011),
www.world-lecture-series.org/san-marino-2011-exp-conf-santilli-isoredshift-iso blueshift

[159] H. Ahmar, G. Amato, J. V. Kadeisvili, J. Manuel, G. West, and O. Zogorodnia, "Additional experimental confirmations of Santilli's IsoRedShift and the consequential lack of expansion of the universe,"Journal of Computational Methods in Sciences and Engineering, **13**, 321 (2013),
http://www.santilli-foundation.org/docs/IRS-confirmations-212.pdf

[160] F. Zwicky, "On the Red Shift of Spectral Lines through Interstellar Space," Proceedings of the National Academy of Sciences of the United States of America, **15**, 773 (1929),
http://www.pnas.org/cgi/reprintframed/15/10/773

[161] P. A. LaViolette, "Is the universe really expanding?", The Astrophysical

Journal **301**, 544 (1986).

[162] R. M. Santilli, "Representation of galactic dynamics via isoshifts without universe expansion, dark matter and dark energy," American Journal of Modern Physics **4**, 26, 2015,
http://www.thunder-energies.com/docs/dark-matter-2015.pdf

[163] R. M. Santilli, "Cracks in the 20th century cosmology," in press,
http://www.santilli-foundation.org/docs/cosmology-notes-2021.pdf

[164] P. Fleming, "Collected papers, interviews, seminars and international press releases on the lack of expansion of the universe,"
http://www.santilli-foundation.org/docs/No-universe-expans.pdf

[165] C. R. Illert and R. M. Santilli, *Foundations of Theoretical Conciology,* Hadronic Press (1995)
http://www.santilli-foundation.org/docs/santilli-109.pdf

[166] D. I. Blochintsev, Phys. Rev. Lett. **12**, 272 (1964).

[167] A. K. Aringazin, Hadronic J. **12** 71 (1989).

[168] S. H. Aronson et al., Phys. Rev. D **28**, 495 (1983).

[169] N. Grossman et al., Phys. Rev. Lett. **59**, 18 (1987).

[170] F. Cardone, R. Mignani and R. M. Santilli "Lie-isotopic energy dependence of the KS lifetime," J. Phys. G: Nucl. Part. Phys. **18**, L141 (1992).

[171] F. Cardone and R. Mignani, Nonlocal approach to the Bose-Einstein correlation, Univ. of Rome Preprint No. 894, July 1992.

[172] Yu. Arestov, R. M. Santilli and V. Solovianov, "Experimental evidence on the isominkowskian character of the hadronic structure," Foundation of Physics Letters **11**, 483 (1998),
http://www.santilli-foundation.org/docs/Santilli-52.pdf

[173] R. Weiner,*Introduction to Bose-Einstein correlations and subatomic inter-*

ferometry, Chichester, England New York: John Wiley (2000).

[174] C. S. Burande, "Bose-Einstein Correlation within the Framework of Hadronic Mechanics," AIP Conference Proceedings 1648, 510007; doi: 10.1063/1.4912712 (2015),
www.santilli-foundation.org/docs/1.4912712(CS-Burande(2)).pdf

[175] R. M. Santilli, "Universality of special isorelativity for the invariant description of arbitrary speeds of light," Nova collection of papers dedicated to the Lorentz symmetry, V. Dvoeglazov, Editor, Nova Publisher (1997), arXiv:physics/9812052 (1997)
https://arxiv.org/pdf/physics/9812052.pdf

[176] R. M. Santilli, "Compatibility of Arbitrary Speeds with Special Relativity Axioms for Interior Dynamical Problems," American Journal of Modern Physics, **5**, 143 (2016),
http://www.santilli-foundation.org/docs/ArbitrarySpeeds.pdf

[177] R. M. Santilli, "Isotopic quantization of gravity and its universal isopoincare' symmetry" in the *Proceedings of "The Seventh Marcel Grossmann Meeting in Gravitation,* SLAC 1992, R. T. Jantzen, G. M. Keiser and R. Ruffini, Editors, World Scientific Publishers pages 500-505 (1994),
www.santilli-foundation.org/docs/Santilli-120.pdf

[178] R. M. Santilli, "Unification of gravitation and electroweak interactions" in the proceedings of the Eight Marcel Grossmann Meeting in Gravitation, Israel 1997, T. Piran and R. Ruffini, Editors, World Scientific, pages 473-475 (1999),
www.santilli-foundation.org/docs/Santilli-137.pdf

[179] R. M. Santilli, "Isominkowskian unification of special and general relativities." arXiv:physics/9705015 (1997)
https://arxiv.org/pdf/physics/9705015.pdf
"Classical and operator isominkowskian unification of general and special relativity for matter and their isodual for antimatter," arXiv:physics/9705016 (1997)
https://arxiv.org/pdf/physics/9705016.pdf (1997),

"Isominkowskian formulation of gravity," arXiv:physics/9707018 (1997) https://arxiv.org/pdf/physics/9707018.pdf

[180] R. M. Santilli, "Rudiments of IsoGravitation for Matter and its IsoDual for AntiMatter," American Journal of Modern Physics **4**, 59 (2015), www.santilli-foundation.org/docs/10.11648.j.ajmp.s.2015040501.18.pdf

[181] R. M. Santilli, "Open problems in GRT and their possible Resolutions via Isogravitation," Galilean Electrodynamics, Summer 2006, **17**, 43 (2006), http://www.i-b-r.org/Incons.GravFinalGED-I.pdf

[182] Debate on gravitation, http://eprdebates.org/general-relativity.php

[183] Schwartzschild K., " Uber das Gravitationsfeld eines Massenpunktesnack der Einsteinshen Theorie, " Sitzber. Deut. Akad. Wiss. Berlin, Kl. Math.-Phys. Tech., 189-196 (1916).

[184] Schwartzschild K., "Uber das Gravitationsfeld einer Kugel aus inkompressibler Flussigkeit nach Einsteinshen Theorie, " Sitzber. Deut. Akad. Wiss. Berlin, Kl. Math. - Phys. Tech., 424-434 (1915).

[185] R. M. Santilli, "Partons and Gravitation: some Puzzling Questions," (MIT) Annals of Physics, **83**, 108 (1974), http://www.santilli-foundation.org/docs/Santilli-14.pdf

[186] R. Mignani, Lett. Nuovo Cimento **39**, 406 (1984); Lett. Nuovo Cimento **43**, 355 (1985); Hadronic Journal **9**, 103 (1986).

[187] A. O. E. Animalu and R. M. Santilli, "Nonunitary Lie-isotopic and Lie-admissible scattering theories of hadronic mechanics," in the *Proceedings of the Third International Conference on the Lie-Admissible Treatment of Irreversible Processes*, Kathmandu University, page 165 on (2011):
Paper I: Conceptual and Mathematical foundation
www.santilli-foundation.org/docs/Isoscattering-I.pdf
Paper II: Deformations-isotopies of Lie's theory, special relativity and mechanics

www.santilli-foundation.org/docs/Isoscattering-II.pdf
Paper III: Basic Lie-isotopic formulations without divergencies
www.santilli-foundation.org/docs/Isoscattering-III.pdf
Pawer IV: Reversible Electron-Proton and Electron-Positron Scatterings
www.santilli-foundation.org/docs/Isoscattering-IV.pdf
Paper V: Foundations of the Genoscattering Theory for Irreversible Processes
www.santilli-foundation.org/docs/Isoscattering-V.pdf

[188] Proceedings of the Second Workshop on Lie-Admissible formulations, Part A: Review Papers, Hadronic Journal . **2**, No. 6 (1979), http://www.santilli-foundation.org/docs/hj-2-6-1979.pdf

[189] Proceedings of the Second Workshop on Lie-Admissible formulations, Part B: Research Papers, Hadronic Journal **3**, No. 1 (1979), http://www.santilli-foundation.org/docs/hj-3-1-1979.pdf

[190] Proceedings of the Third Workshop on Lie-Admissible Formulations, Part A: Mathematics, Hadronic Journal **4**, No. 2 (1981), http://www.santilli-foundation.org/docs/hj-4-2-1981.pdf

[191] Proceedings of the Third Workshop on Lie-Admissible Formulations, Part B: Theoretical Physics, Hadronic Journal **4**, No. 3 (1981), http://www.santilli-foundation.org/docs/hj-4-3-1981.pdf

[192] Proceedings of the Third Workshop on Lie-Admissible Formulations, Part C: Experimental Physics and Bibliography, Hadronic Journal **4**, No. 4 (1981), http://www.santilli-foundation.org/docs/hj-4-4-1981.pdf

[193] Proceedings of the First International Conference on Nonpotential Interactions and their Lie-Admissible Treatment, Part A: Invited Papers, Hadronic Journal **5**, No. 2 (1982), http://www.santilli-foundation.org/docs/hj-5-2-1982.pdf

[194] Proceedings of the First International Conference on Nonpotential Interactions and their Lie-Admissible Treatment, Part B: Invited Papers, Hadronic Journal **5**, No. 3 (1982),

http://www.santilli-foundation.org/docs/hj-5-3-1982.pdf

[195] Proceedings of the First International Conference on Nonpotential Interactions and their Lie-Admissible Treatment, Part C: Contributed Papers, Hadronic Journal **5**, No. 4 (1982),
http://www.santilli-foundation.org/docs/hj-8-4-1982.pdf

[196] Proceedings of the First International Conference on Nonpotential Interactions and their Lie-Admissible Treatment, Part D: Contributed Papers, Hadronic Journal **5**, No. 5 (1982),
http://www.santilli-foundation.org/docs/hj-5-5-1982.pdf

[197] Proceedings of the First Workshop on Hadronic Mechanics, Hadronic Journal **6**, No. 6 (1983),
http://www.santilli-foundation.org/docs/hj-6-6-1983.pdf

[199] Proceedings of the Second Workshop on Hadronic Mechanics, Volume I, Hadronic Journal **7**, No. 5 (1984),
http://www.santilli-foundation.org/docs/hj-7-5-1984.pdf

[199] Proceedings of the Second Workshop on Hadronic Mechanics, Volume II, Hadronic Journal **7** , No. 6 (1984),
http://www.santilli-foundation.org/docs/hj-7-6-1984.pdf

[200] Proceedings of the third international conference on the Lie-admissible treatment of non-potential interactions Kathmandu University, Nepal, 2011, Volume I:
http://www.santilli-foundation.org/docs/2011-nepal-conference-vol-1.pdf Volume II:
http://www.santilli-foundation.org/docs/2011-nepal-conference-vol-2.pdf

[201] D. S. Sourlas and G. T. Tsagas, Mathematical Foundation of the Lie-Santilli Theory, Ukraine Academy of Sciences (1993),
http://www.santilli-foundation.org/docs/santilli-70.pdf

[202] J. Lohmus, E. Paal, and L. Sorgsepp, *Non-associative Algebras in Physics,*

Hadronic Press, Palm Harbor, (1994),
www.santilli- foundation.org/docs/Lohmus.pdf
[203] J. V. Kadeisvili, Santilli's Isotopies of Contemporary Algebras, Geometries and Relativities, Ukraine Academy of Sciences, Second edition (1997),
http://www.santilli-foundation.org/docs/Santilli-60.pdf

[204] Chun-Xuan Jiang, *Foundations of Santilli Isonumber Theory,* International Academic Press (2001),
http://www.i-b-r.org/docs/jiang.pdf

[205] Raul M. Falcon Ganfornina and Juan Valdes, *Fundamentos de la Isoteoria de Lie-Santilli,* International Academic Press (2001),
http://www.i-b-r.org/docs/spanish.pdf

[206] I. Gandzha and J Kadeisvili, *New Sciences for a New Era: Mathematical, Physical and Chemical Discoveries of Ruggero Maria Santilli,* Sankata Printing Press, Nepal (2011),
http://www.santilli-foundation.org/docs/RMS.pdf

[207] S. Georgiev, *Foundation of the IsoDifferential Calculus,* Volume I to VI (2014).

[208] A first tribute to Albert Einstein: Debate on the physical applications of the EPR argument in physics,
http://eprdebates.org/santilli-confirmation-of-the-epr-argument.php

[209] A second tribute to Albert Einstein: Debate on the applications of the EPR argument in chemistry,
http://eprdebates.org/santilli-confirmation-of-the-epr-argument-chemistry.php

PROCEEDINGS OF THE 2020 EPR TELECONFERENCE
Curran Associates Conference Proceedings, New York, USA (2021)

PART I: REPRINTED PAPERS

[210] R. M. Santilli, "Isorepresentation of the Lie-isotopic SU(2) Algebra with Application to Nuclear Physics and Local Realism,"
Reprinted from Acta Applicandae Mathematicae **50**, 177 (1998),
Proceedings of the 2020 Teleconference on the EPR argument,
Curran Associates Conference Proceedings, New York, USA), 171-184 (2021).

[211] R. M. Santilli, "Studies on the classical determinism predicted by A. Einstein, B. Podolsky and N. Rosen,"
Reprinted from Ratio Mathematica **37**, 5 (2019),
Proceedings of the 2020 Teleconference on the EPR argument,
Curran Associates Conference Proceedings, New York, USA, 185-203 (2021).

[212] R.M. Santilli, "Studies on A. Einstein, B. Podolsky, and N. Rosen prediction that quantum mechanics is not a complete theory, paper I: Basic methods,"
Reprinted from Ratio Mathematica **38**, 5, (2020)
Proceedings of the 2020 Teleconference on the EPR argument,
Curran Associates Conference Proceedings, New York, USA, 205-209 (2021).

[213] R.M. Santilli, "Studies on A. Einstein, B. Podolsky, and N. Rosen prediction that quantum mechanics is not a complete theory, paper II: Apparent proof of the EPR argument,"
Reprinted from Ratio Mathematica **38**, 71 (2020),
Proceedings of the 2020 Teleconference on the EPR argument,
Curran Associates Conference Proceedings, New York, USA, 271-338 (2021).

[214] R.M. Santilli, "Studies on A. Einstein, B. Podolsky, and N. Rosen prediction that quantum mechanics is not a complete theory, paper III: Illustrative examples and applications,"
Reprinted from Ratio Mathematica **38**, 139 (2020),
Proceedings of the 2020 Teleconference on the EPR argument,

Curran Associates Conference Proceedings, New York, USA, 339-422 (2021).

PART II: PROCEEDINGS PAPERS

[215] A. O. E. Animalu and R. M. Santilli, "Nonunitary Lie-Isotopic and Lie-Admissible Scattering Theories of Hadronic Mechanics: Irreversible Deep-Inelastic Electron-Positron and Electron-Proton Scattering,"
Proceedings of the 2020 Teleconference on the EPR argument,
Curran Associates Conference Proceedings, New York, USA, 423-458 (2021).

[216] S. Beghella Bartoli, "Significance for the EPR Argument of the Neutron Synthesis from Hydrogen and of a New Controlled Nuclear Fusion without Coulomb Barrier,"
Proceedings of the 2020 Teleconference on the EPR argument,
Curran Associates Conference Proceedings, New York, USA, 459-466 (2021).

[217] E. T. D. Boney, "Completeness is Unfalsifiable: Gödel and Popper for the EPR Debate / Kuhn and the Standard Model,"
Proceedings of the 2020 Teleconference on the EPR argument,
Curran Associates Conference Proceedings, New York, USA, 467-480 (2021).

[218] Y. F. Chang, "Nonlocality, Entangled Field and IIS Predictions, Superluminal Communication,"
Proceedings of the 2020 Teleconference on the EPR argument,
Curran Associates Conference Proceedings, New York, USA, 481-494 (2021).

[219] I. B. Das Sarma, "Comparison of Various Nuclear Fusion Reactions and ICNF,"
Proceedings of the 2020 Teleconference on the EPR argument,
Curran Associates Conference Proceedings, New York, USA, 495-502 (2021).

[220] J. Dunning-Davies, "Inaugural Lecture,"
Proceedings of the 2020 Teleconference on the EPR argument,
Curran Associates Conference Proceedings, New York, USA, 503-513 (2021).

[221] T. Ernst Zur Theorie, "Der q,w-Lieschen Matrixgruppen," from

Proceedings of the 2020 Teleconference on the EPR argument,
Curran Associates Conference Proceedings, New York, USA, 515-537 (2021).

[223] S. G. Georgiev, "Bi-α Iso-Differential Calculus,"
Proceedings of the 2020 Teleconference on the EPR argument,
Curran Associates Conference Proceedings, New York, USA, 539-557 (2021).

[222] S. G. Georgiev,"Bi-α Iso-Differential Inequalities and Applications,"
Proceedings of the 2020 Teleconference on the EPR argument,
Curran Associates Conference Proceedings, New York, USA, 559-605 (2021).

[224] S. E. Johansen, "Principle(s) of Causality as De Facto Fundamental in Mathematical Physics, Including For a Chance Causality Applied in Quantum Mechanics,"
Proceedings of the 2020 Teleconference on the EPR argument,
Curran Associates Conference Proceedings, New York, USA, 607-643 (2021).

[225] S. Kumar, "A General Relativistic Theory of Electromagnetic Field and its Connection with Planek's Constant,"
Proceedings of the 2020 Teleconference on the EPR argument,
Curran Associates Conference Proceedings, New York, USA, 645-667 (2021).

[226] A. S. Muktibodh, "Role of the Lie-Santilli Isotheory for the Proof of the EPR Argument,"
Proceedings of the 2020 Teleconference on the EPR argument,
Curran Associates Conference Proceedings, New York, USA, 669-681 (2021).

[227] A. S. Muktibodh, "Isodual Mathematics for Antimatter,"
Proceedings of the 2020 Teleconference on the EPR argument,
Curran Associates Conference Proceedings, New York, USA, 683-700 (2021).

[228] A. A. Nassikas, "Minimum Contradictions Theory of Everything,"
Proceedings of the 2020 Teleconference on the EPR argument,
Curran Associates Conference Proceedings, New York, USA, 701-714 (2021).

[229] R. L. Norman and J. Dunning-Davies, "A Proposed Physical Basis for

Quantum Uncertainty Effects,"
Proceedings of the 2020 Teleconference on the EPR argument,
Curran Associates Conference Proceedings, New York, USA, 715-728 (2021).

[230] O. A. Olkhov, "Possibility of Geometrical Interpretation of Quantum Mechanics and Geometrical Meaning of 'Hidden Variables'"
Proceedings of the 2020 Teleconference on the EPR argument,
Curran Associates Conference Proceedings, New York, USA, 729-739 (2021).

[231] R. M. Santilli and T. Vougiouklis, "A New Conception of Living Organisms and its Representation via Lie-Admissible H_v-Hyperstructures,"
Proceedings of the 2020 Teleconference on the EPR argument,
Curran Associates Conference Proceedings, New York, USA, 741-764 (2021).

[232] E. Trell, "Entanglement is Real in 3-D 'Game of Lie' Straight Line Geometric Algebra Cellular Automaton,"
Proceedings of the 2020 Teleconference on the EPR argument,
Curran Associates Conference Proceedings, New York, USA, 765-781 (2021).

[233] T. Vougiouklis, "Extending Mathematical Models from Numbers to H_v-Numbers,"
Proceedings of the 2020 Teleconference on the EPR argument,
Curran Associates Conference Proceedings, New York, USA, 783-799 (2021).

PART III: PAPERS ADDED IN PROOF

[234] U. V. S. Seshavatharam and S. Lakshminarayana, "EPR argument and the mystery of the reduced Planck's constant."
Proceedings of the 2020 Teleconference on the EPR argument,
Curran Associates Conference Proceedings, New York, USA, 801-821 (2021).

[235] G. Eigen, "Measurements of the Polarization Correlation of the Two-Photon System Produced in Positron-Electron Annihilation"
Proceedings of the 2020 Teleconference on the EPR argument,
Curran Associates Conference Proceedings, New York, USA, 823-843 (2021).

Index

ABOUT THIS BOOK

This book presents an overview of the various topics studied at the 2020 International Teleconference on the Einstein-Podolsky-Rosen Argument that *Quantum mechanics is not a complete theory*, including:

1) The 1935 EPR objections against **quantum entanglements** due to the strictly local character of quantum mechanics with consequential need for superluminal communications;

2) The representation of contact, zero-range, non-linear, non-local and non-Hamiltonian interactions between the overlapping wavepackets of extended particles via the isotopic element $\hat{T}$ of the axiom-preserving isoproduct $A\hat{\times}B = A\hat{T}B$ of isomathematics and related hadronic mechanics;

3) Proof that the Copenhagen interpretation of quantum mechanics is based on the simplest possible realization of quantum axioms (such as the historical product $AB = A \times B$ of the enveloping associative algebra), while hadronic mechanics is based on the most general possible realizations of the same axioms (such as the isoassociative isoproduct $A\hat{\times}B$), to such as extent that the two formulations can be expressed with the same symbols solely differentiated by their realizations;

4) The new type of particle entanglements, here called **EPR entanglements**, consisting of particles in continuous and instantaneous communication via the overlapping of their wavepackets represented by the isotopic element $\hat{T}$ without any need for superluminal communications;

5) Proofs that the isotopic element $\hat{T}$ provides an explicit and concrete realization of Bohm's hidden variables with ensuing inapplicability of Bell's inequality;

6) Historical verifications of the EPR argument via the non-linear theory by W. Heisenberg, the non-local theory by L. de Broglie, D. Bohm's theory of 'hidden variables' and others;

7) Recent verifications of the EPR argument by R. M. Santilli showing the existence of classical images for extended particles under non-Hamiltonian interactions, as well as the progressive recovering of Einstein's determinism in the structure of hadrons, nuclei and stars, and its full recovering at the limit of Schwartzchild's horizon;

8) Proofs that, due to its strict time reversibility, the Copenhagen interpretation of quantum mechanics cannot consistently represent energy-releasing processes such as nuclear fusions or fossil fuel combustion, due to their irreversibility over time;

9) Novel advances in physics, chemistry and biology permitted by reversible and irreversible branches of hadronic mechanics, including the exact representation of nuclear data with ensuing new formulations of controlled nuclear fusions, an attractive force between valence electron bonds with ensuing exact representation of molecular data, and a new conception of life consisting of extended constituents under continuous EPR entanglement represented via hyperstructures.

Biographical notes of
SIR RUGGERO MARIA SANTILLI

Tel. +1-727 688 3992, Email research@i-b-r.org

Academic and scientific notes Ruggero Maria Santilli received the highest possible education in Italy, emigrated in the U.S.A. with his family in 1967 following an invitation from the University of Miami, Florida, to conduct research under NASA support, after which he was in the faculty of Boston University, MIT, and Harvard University under support from NASA, USAFOSR and DOE. From 1985 on, Dr. Santilli has been Professor of Physics and President of The Institute for Basic Research originally located within the compound of Harvard University, became a U. S. Citizen in 1986, and moved to Florida in 1989. Santilli is the author of 325 papers in mathematics, physics and chemistry published in refereed journals, has written 20 Ph. D. level monographs in various fields, is the founder of three scientific journals and the editor of various journals. For more details, please visit the full-length curriculum http://www.i-b-r.org/Ruggero-Maria-Santilli.htm

Corporate notes Santilli has been Scientific Advisor to various U. S. corporations. From 2007 to 2013, Santilli has been Chief Scientist and Chairman of the Board of Magnegas Corporation, a U. S. company publicly traded at NASDAQ under the stock symbol MNGA, producing and selling the gaseous magnegas fuel synthesized from liquid wastes with complete combustion. From 2014 to 2018, Santilli has been the CEO and Chief Scientist of Thunder Energies Corporation, also a publicly traded company with stock symbol TNRG. Santilli is currently the Chief Scientist of Hadronic Technologies Corporation, a U. S. company developing cutting edge new technologies, such as: the synthesis of neutrons from a hydrogen gas and its application; a new combustion of fossil fuels with complete combustion; and new telescopes for the detection of antimatter asteroids, cosmic rays and galaxies. For more details, please visit http://www.hadronictechnologies.com

Honors Ruggero Maria Santilli was knighted with the title of Sir on September 6, 2011 by the Republic of San Marino as a member of the millenary Equestrian Order of Sant'Agata. Sir R. M. Santilli has been the recipient of the following additional honors: the 1982 gold medal for scientific merits from the Université d'Orleans, France; the 1990 nomination by the Estonia Academy of Sciences "among the most illustrious applied mathematicians of all times"; the 2009 Mediterranean Prize; the 2009 scientific prize from the U. S. Sons of Italy; the 2011 scientific prize from Kathmandu University, Nepal; the 2011 membership of the European Society of Computational Methods; the 2015 ICNPAA award at the University of La Rochelle, France; the 2016 Fray International Sustainability Award, granted at the SIPS International Conference, Hainan Island, China. In September 2011; and the recipient on May 31, 2018, of the "Star of Italy" by President Sergio Mattarella. Sir Santilli has been nominated for the Nobel Prize in Physics and, separately, Chemistry since 1987. For more details, please visit the website
http://santilli-foundation.org/santilli-nobel-nominations.html.
July 15, 2021

Hadronic Press
35246 US 19 North Suite 215
Palm Harbor, Florida 34684 USA

ISBN 978-1-7138-6658-9

CURRAN ASSOCIATES INC.
proceedings.com